A Citizen's Guide
to Promoting
Toxic Waste Reduction

Lauren Kenworthy

INFORM, Inc.
381 Park Avenue South
New York, NY 10016
(212) 689-4040

Library of Congress Cataloging-in-Publication Data

Kenworthy, Lauren, 1962-
    A citizen's guide to promoting toxic waste reduction / Lauren
Kenworthy and Eric Schaeffer.
        p.   cm.
    Includes bibliographical references.
    ISBN 0-918780-54-3 : $15.00
    1. Hazardous wastes--Citizen participation. 2. Factory and trade
participation. I. Schaeffer, Eric. II. Title.
TD1050.C57K46   1990
363.72'8757--dc20   89-71650
                                                      CIP

INFORM, Inc., founded in 1974, is a nonprofit research and education organization that identifies and reports on practical actions for the protection and conservation of natural resources and public health. INFORM's research is published in books, abstracts, newsletters and articles. Its work is supported by contributions from individuals and corporations and by grants from over 40 foundations.

Book design by Garret Schenck
Cover design by Saul Lambert

Third Printing

 printed on
recycled paper

Contents

Chapter 1   Introduction   1
    The Need for Reducing Toxic Waste at the Source   1
    The Role You Can Play: Promoting Source Reduction   2
    How This Guide Will Help You   2

Chapter 2   The Source Reduction Option   4
    Defining Source Reduction   4
    Why Is Source Reduction the Best Strategy?   4
    Resources   8

Chapter 3   Source Reduction Strategies Plants Can Use   10
    Five Techniques   10
    Resources   11

Chapter 4   Creating Effective Source Reduction Programs   12
    Company Atmosphere   12
    Five Steps Companies Can Take   13
    Resources   17

Chapter 5   Surveying Your Local Plant   19
    Considering Plants To Study   19
    Conducting Background Research   20
    Preparing for the Plant Interview   30
    Interviewing Plant Representatives   36
    Drawing Conclusions and Using Them   48
    Resources   49

Chapter 6   Government Policies That Encourage Source Reduction   50
    Establishing Source Reduction as the Top Priority   50
    Establishing a State Source Reduction Office   51
    Delegating Nine Responsibilities to the Office   52
    Establishing "Right to Know" Policies   53
    Ensuring All Regulatory Programs Promote Source Reduction   54
    Establishing State Programs Supporting Source Reduction Policies   55
    Resources   56

Appendix A   The Toxics Release Inventory   57
    TRI in a Nutshell   57
    Which Companies Must Report?   57
    What Information Is Reported?   67
    How To Get TRI Information   67
    How To Use TRI Information   73
    What Information TRI Does Not Provide   74
    Resources   75

Appendix B   State Contacts for TRI Information   79

Appendix C   Overview of Known Health Effects of TRI Chemicals   82

Appendix D   Other Useful Right to Know Information   95
    Resources   95

Appendix E   Federal Waste Minimization Forms   98

Appendix F   Glossary   102

Appendix G   Worksheets   105
    Profile of the plant (Worksheet 1)   106
    Plant permit compliance and applications for new permits (Worksheet 2)   107
    Plant pollution profile (Worksheet 3)   109
    Measuring the corporate commitment to source reduction (Worksheet 4)   110
    Waste-related data collection (Worksheet 5)   112
    Measuring company accomplishments (Worksheet 6)   116
    Corporate incentives and barriers to source reduction (Worksheet 7 — optional)   117

Recent INFORM Publications   119

About the Authors   121

INFORM's Board of Directors   122

Preface

In 1989, for the first time ever, the United States Environmental Protection Agency produced the Toxics Release Inventory, a national database on discharges of individual toxic chemicals from individual industrial plants to individual segments of the environment. In 1987 alone, the only year for which figures are now available, 19,278 industrial plants released more than 22.5 billion pounds of 308 specific toxic chemicals (and still more chemicals grouped into 20 categories) to the air, surface waters, sewage treatment plants, landfills, underground injection wells, and off-site facilities. Pollution management costs for government and industry came to more than $85 billion in 1988.

What can individuals constructively do to help stem this tide of environmental pollution, particularly from industrial plants in their communities? At INFORM, we have often been asked this question working with groups around the country and, in response, we are now especially pleased to publish this "how-to" manual for citizens' groups.

In this guide, we have drawn on INFORM's expertise in hazardous waste reduction developed over the last seven years. During this time, we have published *Cutting Chemical Wastes*, an in-depth study of waste reduction initiatives in 29 of this country's chemical plants, and a further paper on steps government can take to promote source reduction.

With *A Citizen's Guide to Promoting Toxic Waste Reduction*, we have sought to share our research methods. Using it, concerned individuals can learn how to use the Toxics Release Inventory and other data to understand the operations and discharges of local plants. They will find step-by-step methods for building positive communications with plant officials and for gathering the information they need to assess source reduction efforts. The author, Lauren Kenworthy, clearly identifies source reduction strategies plants can use and provides resource lists for additional information. The guide also contains a chapter about promoting source reduction through statewide legislation.

Ohio Citizens Action called a draft of this manual "a superb tool." We hope other groups around the country will find it equally useful.

Joanna D. Underwood
President, INFORM

Acknowledgments

We could not have written this guide without the help of the many people who shared ideas with us and reviewed our manuscript at various stages. In particular, we are grateful to David Allen of the National Toxics Campaign, Patricia Bauman of the Bauman Family Foundation, Sandy Buchanan of Ohio Citizens Action, Will Collette of the Citizen's Clearinghouse for Hazardous Waste, Patricia Costner of Greenpeace, Jerry Kotas of EPA's Office of Pollution Prevention, Professor Francis Lynn of the Institute for Environmental Science at the University of North Carolina, Jerry Martin of Dow Chemical Company, David Sarokin of the EPA's Office of Toxic Substances, Ted Smith of the Silicon Valley Toxics Coalition, and Rob Stewart of New Jersey Public Interest Research Group.

We are also extremely grateful to the staff of INFORM, who made this guide a reality, especially to Mark Dorfman, Associate Director of the Chemical Hazards Prevention Program, who shepherded the project through the many phases of writing, reviewing, and production. Joanna Underwood, INFORM's President, provided support and encouragement from the very beginning of the project, as did Nancy Lilienthal, Director of the Chemical Hazards Prevention Program, who also made very helpful suggestions on the manuscript. Thanks go to Sibyl Golden, Senior Editor, whose editing made our final manuscript more readable and usable, and to Production Coordinator Garret Schenck who created the design and layout for this guide and turned the manuscript into a finished product. We also especially appreciate the hard work of our typist, Karen S. Butler, who typed and retyped many drafts of the manuscript.

Finally, we are indebted to the Fund for the City of New York's Nonprofit Computer Exchange, which provided essential production assistance, and to the Bauman Family Foundation, whose generous grant covered most of the costs of this project. Supplementary support came from grant funds contributed to INFORM's Chemical Hazards Prevention Program by the Geraldine R. Dodge, Joyce, and Charles Stewart Mott Foundations, and by the Fund for New Jersey.

Although this book benefitted from the advice of many individuals, the contents remain the responsibility of the authors and INFORM.

Chapter 1

Introduction

Since World War II, chemicals have played an increasingly large role in our daily life. Almost without becoming aware of it, we have grown dependent on medicines, plastics, automobiles, synthetic textiles, and a myriad of other products as part of our standard of living. Chemistry has enabled us to live longer and more comfortably, surrounded by many conveniences that we take for granted.

"Better living through chemistry" has come with a price, however. In 1987 alone, the only year for which such data are now available, 19,278 United States industrial plants poured more than 22.5 billion pounds of toxic chemicals into our air, land, and water — and these figures include only 308 individual chemicals and 20 chemical groups out of the 70,000 in commercial use today. Beginning with Rachel Carson, environmentalists have alerted us to the irreparable damage such chemical pollution can cause to our environment.

The Need for Reducing Toxic Waste at the Source

State and federal governments have responded to this awakening of concern with passage of a growing number of environmental laws. These laws have been strict in some areas (we have banned outright certain types of disposal on land), but much less strict in others (federal law regulates only seven toxic air pollutants). These regulations need to be maintained and in some areas strengthened. But, for several reasons, we cannot rely solely on pollution control regulations for environmental protection:

- Many pollutants remain unstudied, unmonitored, and unregulated. Even in the area of land disposal, which has received intensive attention from regulators, the United States General Accounting Office found that the Environmental Protection Agency (EPA) "does not know whether it is controlling 90% of existing hazardous waste or 10%" (GAO, *New Approaches Needed to Manage RCRA*, RCED-88-115, 1988, p.19).

- Many regulations are technically complex and can take years, or even decades, to implement. Some become the subject of protracted legal battles, lengthening the process still further.

- Officials charged with granting environmental permits and enforcing environmental regulations are stretched to the breaking point on the federal and state levels. Lean budgets limit funds for their activities, and large budget deficits foreclose hopes for the major spending increases needed to turn this situation around.

Just as regulators are being overwhelmed by the quantities of pollution facing this country, the public is coming to understand that no matter how many good pollution control laws we pass, the planet is simply too small to handle the volumes dumped onto it each year. Pollution control systems have often simply ended up moving pollutants from one environmental medium to another rather than actually getting rid of them. "Miracle machines" like incinerators, for example, consume wastes from land dumps, often only to turn them into toxic air pollutants, while smokestack scrubbers that capture air pollutants often create more toxic solid waste in the form of ash. We can never win this "toxic shell game."

At the same time, industry has awakened to the fact that the liability for cleaning up old pollutants, and the cost of managing new pollution, adds up to an enormously expensive problem. Of the over $85 billion a year now spent on pollution control by the United States government and business, business foots two-thirds of the bill.

Source reduction of wastes is a strategy that addresses these problems by reducing the production of **hazardous** and **toxic wastes** in the first place, instead of cleaning them up after they are created. Like energy efficiency, it is a strategy that protects the environment and often saves companies money at the same time. By actively seeking out ways to avoid producing wastes in the first place, companies can operate more efficiently, wasting fewer raw materials, while they create less pollution and reduce their pollution control, handling, and liability costs.

The Role You Can Play: Promoting Source Reduction

Your help is needed. It is time for citizens to challenge industry to adopt a "good neighbor" policy and work with local communities to remedy those pollution problems that environmental laws and regulations have not yet solved. Source reduction is the ideal tool for this task. But source reduction is a relatively new idea for industry, and in many cases, businesses may not fully explore source reduction opportunities without some respectful prodding from people like you. That is why your involvement is essential.

How This Guide Will Help You

This guide will help you study and evaluate the source reduction efforts of factories in your community. With it, you can actively promote source reduction at local plants and encourage the reduction or elimination of potentially dangerous chemical discharges, whether or not they are currently regulated by law. The guide equips you with the arguments you need to convince businesses in your neighborhood that source reduction is the environmental protection method of choice and should be the first line of defense against pollution. In the long run, this guide can help you help local businesses to become better, cleaner neighbors.

Chapters 2-5 of this guide provide you with the tools you need to promote source reduction at individual plants in your community.

- Chapter 2 introduces you to the concept of source reduction, describes its benefits, and explains how it is fundamentally different from other pollution control methods.

- Chapters 3 and 4 describe the strategies plants can use to reduce pollution and the steps company management should take to create an effective source reduction program.

- Chapter 5 gives you a step-by-step research process to follow when studying local plants, including questions for an interview with company representatives.

- Each chapter steers you to experts and publications that provide more detailed facts and different perspectives on source reduction. (Each section of the guide is followed by a list of key sources and important documents.)

By following the steps for researching and evaluating local plants, you will be able to:

- *create a dialogue* with company representatives and spur the company to explore source reduction opportunities;

- *assess whether a company has created the policies*, management incentives, and informational tools required for a serious source reduction program;

- *identify specific source reduction actions* taken at a given company and draw basic conclusions about the success of the company program over time.

This guide will *not* enable you to judge the technical aspects of specific source reduction actions taken because this requires detailed knowledge and understanding of the individual production processes involved. However, it *will* allow you to judge the overall source reduction program and send a powerful message that you are expecting a responsible, documentable good neighbor policy from local companies when it comes to preventing pollution with source reduction.

Finally, Chapter 6, to help you promote source reduction, outlines some of the legislative approaches that can support your efforts at the local level by enabling you to get more information on local plants and their pollution records and by encouraging industrial source reduction through new incentives and requirements.

Additionally, appendices provide details on the Toxics Release Inventory (TRI), including information about state contacts and known health effects of TRI chemicals, and on other sources of information that you may find useful as you research plants.

While this guide has been written for readers who do not have a background in industrial chemistry or other technical areas, it has sometimes been necessary to introduce a few technical terms. These terms are highlighted with **boldface** when they are first used, defined in the text, and listed in the Glossary.

Chapter 2

The Source Reduction Option

Source reduction often seems to have a variety of meanings. It is important that you and the industrial representatives you deal with have a firm shared understanding of the term: both what it includes and what it does not include.

Defining Source Reduction

The congressional Office of Technology Assessment (OTA) and INFORM, Inc., have published studies on source reduction with complementary definitions:

- According to OTA, in *Serious Reduction of Hazardous Waste* (see Resources), source reduction covers "in-plant practices that reduce, avoid or eliminate the generation of hazardous waste so as to reduce risks to health and environment." (OTA includes **recycling** that is "an integral part of the . . . industrial process," or "in-process" recycling, in its definition. It does not include recycling that involves moving wastes to another area of the plant or to an off-site location.)

- INFORM, in *Cutting Chemical Wastes* (see Resources), states that source reduction "means that the source of the waste is altered in some way so as to reduce the amount generated or eliminate it altogether."

Both OTA and INFORM limit their definitions of source reduction to *action taken prior to out-of-process recycling, treatment, disposal, or any other form of managing waste.* Both organizations define waste and hazardous waste in broad terms, such as "all nonproduct hazardous outputs . . ." (OTA). Both present source reduction options — some technical and some not — that either increase the efficiency of chemical use in a plant or eliminate the use of a chemical altogether.

You may hear company officials talk about "waste reduction," "waste minimization," or "pollution prevention." These are not necessarily the same as source reduction, and you need to make sure that you and the company are both talking about the same set of strategies. (See box: Why "Source Reduction" Is Not Necessarily "Waste Reduction" or "Pollution Prevention.")

Why Is Source Reduction the Best Strategy?

This guide focuses only on source reduction as it is defined above because only source reduction offers a multitude of environmental and economic benefits.

4

Why "Source Reduction" Is Not Necessarily "Waste Reduction" or "Pollution Prevention"

Various terms that you may hear confuse the concept of source reduction. For example, the terms "waste minimization," and even "waste reduction" may refer to recycling and treatment as well as to true source reduction practices. How can this be so? It is because many companies (and many government agencies) primarily concerned with landfilled wastes consider incineration, biological treatment, or other approaches taken after the creation of waste as measures "reducing" or "minimizing" wastes going to landfill, even though such measures do not actually prevent waste generation in the first place.

You do not want to let a company go unchallenged if it claims that it is preventing pollution because it has a new incinerator on its grounds: it is, in fact, burning waste *after* it has been created. It may be transforming that waste into toxic air emissions that you and your neighbors must breathe, or into ash that must be disposed of in land dumps from which chemicals can seep into groundwater.

Terms like "**waste minimization**" most often signal that a company program (or a federal or state program) is designed to limit land disposal, which may be a good and important waste management step to take under some circumstances. *But* it does not necessarily mean that pollutants are no longer being generated just because they are not going to landfill.

Another term that you may hear is "**pollution prevention.**" The Environmental Protection Agency defines this term to include environmentally sound recycling as well as true source reduction. If someone uses this term, make sure you clarify just what strategies are included.

Source reduction is preventive. By reducing or eliminating toxic waste at its industrial source, source reduction is the only pollution management approach that prevents the creation of waste altogether. The advantages of prevention over **end-of-the-pipe** pollution control strategies, which address the problem *after* it has been created, are obvious. Because source reduction happens before waste is created, managed, handled, or stored, it is the safest way to eliminate pollution threats to workers, the environment, and the public. It reduces the amount of pollutants transported from, spilled at, or emitted at a factory. It prevents the transfer of pollution from one part of the environment to another discussed earlier: transfer that occurs, for example when sludge from air pollution control scrubbers is dumped on the land or toxic chemicals from solid waste seep into groundwater supplies.

Source reduction protects workers, communities, and the environment better than any pollution control technique can. Therefore, it is at the top of the hierarchy for waste management strategies now widely accepted by the United Nations, the United States Environmental Protection Agency, the United States Congress, the Chemical Manufacturers Association, Greenpeace International, and many other organizations. Recycling is considered the next best way to handle waste. Treatment — for example, biological degradation of toxic chemicals, well controlled incineration of **hazardous waste,** or smokestack scrubbers that remove pollutants from air emissions — is next in line. Finally, legally permitted disposal — for example, putting waste-filled barrels in landfills — is the least desirable waste management technique (see box next page: Waste Management Hierarchy).

Waste Management Hierarchy

SOURCE REDUCTION

Reduce/Eliminate the Use of Hazardous or Toxic Materials
- Materials substitution
- Product substitution or reformulation

Reduce/Eliminate the Generation of Wastes
- Technology/equipment modification
- Improved plant operations (In-process recycling)
- Process changes

ON-SITE RECYCLING/RESOURCE RECOVERY

(leaves hazardous or toxic residual for disposal)

*OFF-SITE RECYCLING/RESOURCE RECOVERY

(leaves hazardous or toxic residual for disposal)

ON-SITE TREATMENT

- Physical treatment
- Chemical treatment
- Biological treatment
- Thermal treatment

(leaves hazardous or toxic residual and/or involves air/water release)

*OFF-SITE TREATMENT

- Physical treatment
- Chemical treatment
- Biological treatment
- Thermal treatment

(leaves hazardous or toxic residuals and/or involves air/water release)

*RESIDUAL REPOSITORIES

(Solid residuals from recycling and treatment, disposed of with ground and air cover. A form of land disposal with more protection than today's landfills.)

*Off-site facilities potentially require "siting."

Source reduction is a "multimedia" concept. While tragedies like Love Canal have focused our attention on the threats of land disposal, source reduction attacks pollution from a much broader perspective. A complete source reduction program is **multimedia**: it addresses pollution going to all parts, or media, of the environment — land, water, and air (see box: Source Reduction Applies to *All* Toxic Pollutants). Candidates for source reduction include plant air emissions, whether intentional **point-source releases** or unintentional **fugitive releases**; wastewater discharges; and releases of solid wastes, whether they are sent to incinerators, publicly owned treatment works, other forms of treatment, or landfills. And source reduction strategies can be used regardless of whether plant releases are currently regulated by state or federal laws.

Source Reduction Applies to All *Toxic Pollutants*

Over the years, the federal government has developed several lists of hazardous pollutants, each relating to a different law. Some of these laws group pollutants by the medium — land, air, or water — that is the subject of the regulations. Others stipulate procedures plants must follow if they use certain chemicals. Source reduction targets the chemicals on all of these lists.

- *Clean Air Act Hazardous Air Emissions.* Six specific chemicals (asbestos, beryllium, mercury, vinyl chloride, benzene and arsenic) and one generic category (radionuclides) released into the air.

- *Clean Water Act Priority Pollutants.* One hundred and twenty-six individual chemicals released into the water, including volatile organic substances such as benzene, chloroform, and vinyl chloride; acid compounds such as phenol and its derivatives; pesticides such as chlordane, DDT, and toxaphene; heavy metals such as lead and mercury; PCBs; and other organic and inorganic compounds.

- *Resource Conservation and Recovery Act (RCRA) Hazardous Wastes.* More than 400 discarded commercial chemical products and specific chemical constituents of industrial waste streams destined for treatment or disposal on land.

- *Superfund Amendments and Reauthorization Act (SARA) Title III Section 313 Toxic Substances.* More than 320 chemicals and chemical categories released into all three environmental media; under specified conditions, facilities must report releases of these chemicals to EPA's annual Toxics Release Inventory.

- *Superfund Amendments and Reauthorization Act (SARA) Section 302 Extremely Hazardous Substances.* More than 360 chemicals for which facilities are required to prepare emergency action plans if these chemicals are present at the facility above certain threshold quantities. Releases of these chemicals to the air, land, or water trigger required reporting by the facility to the State Emergency Response Committee (SERC) and the Local Emergency Planning Committee (LEPC) under SARA Section 304.

In fact, source reduction measures may be readily applied to other potentially toxic or hazardous substances not yet singled out for regulation, or to the balance of the 70,000 chemicals in commercial use today whether or not their toxicology is fully understood.

A multimedia source reduction program is essential for two reasons. It *avoids the so-called toxics shell game* through which a program that addresses emissions to only one medium may in fact be shifting pollution to another environmental medium without reducing overall pollution at all. It also *best prevents health hazards*. The EPA and others have documented that pollutants can pose serious health hazards in all parts of the environment. Recent studies indicate that discharges of chemical toxins into the air — discharges that have been the object of the least government and industry attention for many years — may pose some of the most serious health threats of all.

Source reduction is cost-effective. Most pollution management strategies cost a lot of money. While as a nation we spent more than $85 billion in 1988 managing and controlling pollution, source reduction often *saves industry money* by avoiding loss of valuable raw materials and end products that were formerly discharged into the environment and by reducing expensive pollution management costs arising from waste handling, regulatory compliance, and liability.

Source reduction offers "win-win" solutions to environmental problems by accomplishing better environmental protection and saving industry money at the same time. A single Dow Chemical plant, for example, saves over $3.5 million dollars per year through just three source reduction techniques implemented over two years. When you ask company officials to explore source reduction, you are asking them to do themselves a favor.

Source reduction is often easy to carry out. Source reduction techniques include many low- or no-technology approaches that require very little capital to implement.

Source reduction has tremendous untapped potential. INFORM's study of source reduction at 29 United States organic chemical plants, *Cutting Chemical Wastes*, identified significant source reduction achievements in chemical plants of varying sizes and types. Among the 44 specific source reduction initiatives found at these plants, some resulted in the virtual elimination of waste streams, while others resulted in reductions of 30, 50, or 80 percent, or more. The congressional Office of Technology Assessment, through its industry research, estimates serious source reduction efforts could reduce pollution by at least 50 percent across the nation over a five-year period. The Environmental Protection Agency's 1986 estimate of 33 percent is more conservative, but still indicates the potential for source reduction to radically change pollution levels in all our communities.

Resources

INFORM, Inc. (David J. Sarokin; Warren R. Muir, Ph.D.; Catherine G. Miller, Ph.D.; and Sebastian R. Sperber). *Cutting Chemical Wastes.* New York: INFORM, Inc., 1985. (Available from INFORM, Inc., 381 Park Avenue South, New York, NY 10016. (212) 689-4040. Price: $47.50 plus $2.50 handling charge; 25 percent discount for nonprofit organizations.)

INFORM's study of 29 organic chemical plants in Ohio, New Jersey, and California offers a detailed description of the pollutants generated by and source reduction activities at each facility, as well as a general introduction to source reduction. The case studies will give you useful background as you study plants in your area.

National Toxics Campaign. *The Citizens Toxics Protection Manual.* (Available from the National Toxics Campaign, 37 Temple Place, 4th floor, Boston, MA 02111. (617) 482-1477. Price: $35.00.)

The scope of this document is broader than source reduction, but it contains an informative chapter on toxics use reduction as well as excellent corporate research tips.

North Carolina Pollution Prevention Pays Program. *Accomplishments of North Carolina Industries.* (Available from Pollution Prevention Pays Program, North Carolina Department of Natural Resources, PO Box 27687, Raleigh NC 27611. (919) 733-7015. Price: $5.00.)

The North Carolina Pollution Prevention Pays Program, the most well known state technical assistance program in the country, has documented useful case histories and industry guidelines.

Office of Technology Assessment. *Serious Reduction of Hazardous Waste.* (Summary available from the Office of Technology Assessment, United States Congress, Washington, DC 20510. (202) 224-8996. Free. Book available from the National Technical Information Service, 5285 Port Royal Road, Springfield, VA 22161. Price: $28.95 in paperback or $6.95 in microfiche, plus $3.00 per order for handling for either edition.)

The OTA study of source reduction contains information on everything from definition to state programs to examples of source reduction to case histories documenting the failure of regulatory programs. Like *Cutting Chemical Wastes*, it is an excellent comprehensive introduction to the issue.

Chapter 3

Source Reduction Strategies
Plants Can Use

Plant managers who want to explore source reduction opportunities at their plants can examine a multitude of possible strategies. Despite a common assumption that source reduction must necessarily be directed at altering the manufacturing process, INFORM's case study research, and that of others, has found widespread potential for reducing pollution originating from all stages of the handling of hazardous chemicals. Pollutants have been reduced from storage, loading, transfer, and pollution control operations, as well as in the manufacturing process itself.

Five Techniques

The actual source reduction techniques plants can use fall into five broad categories: chemical substitutions, product reformulations, production process changes, equipment changes, and maintenance and housekeeping modifications.

Chemical substitution means using raw materials that create fewer pollutants in the production process without changing the product itself. A common generic form of chemical substitution is the replacement of toxic organic solvents with nontoxic water-based solvents in a variety of production processes. For example, Ilco Unican Corporation in Rocky Mount, North Carolina, eliminated the use of the toxic organic solvent 1,1,1-trichloroethane for degreasing certain of its metal products. The company installed a new system for the degreasing operations that used a water-based biodegradable solution instead of the organic solvent. Savings in 1988 totaled over $40,000.

Product reformulation involves redesigning the end product so its manufacture creates less hazardous pollution. The 3M Company, for example, reformulated a product so that a nonhazardous organic raw material could be used in its manufacture instead of a cadmium alloy, thereby eliminating a cadmium waste stream.

Production process changes can improve efficiency and reduce pollution resulting from production activities. For example, since 1952, Monsanto's plant in Addyston, Ohio, has gradually changed its polystyrene process from one of batch reactions to a closed-system continuous reaction and achieved a 99 percent reduction in air emissions. (**Continuous processes** are dedicated to continuous production

10

of a single product, whereas **batch processes** are used to produce different products at different times to meet changing customer needs. Continuous process operations usually generate less pollution per pound of product than do batch process operations.)

Equipment changes can dramatically reduce pollution. An Exxon plant in Bayway, New Jersey, installed floating roofs on 16 tanks storing the most volatile (easily evaporated) chemicals used at the plant. A floating roof is a cover that rests on the surface of the liquid being stored in the tank. The roof rises and falls as the level of the liquid changes, thus minimizing the formation of chemical vapors that have to be vented from the tank. Emissions reductions in 1983 totalled 680,000 pounds per year, worth more than $200,000.

An important subcategory of equipment changes is in-process recycling. Often, a particular process creates a pollutant that can be fed directly back into the process through a series of closed pipes, thus eliminating the pollutant and increasing the product yield. For example, USS Chemicals in Ironton, Ohio, added resin absorption systems to the cumene oxidation process in its phenol production plant. The absorption systems soak up organic vapors and return them to the production process for use as raw materials. This change has prevented 715,000 pounds of cumene vapor releases each year and saved USS Chemicals $178,750 annually in cumene costs.

Maintenance/housekeeping modifications involve improving plant operations, including material handling and equipment maintenance, in order to create less waste. Borden Chemical Company's Fremont, California, plant, studied by INFORM, reduced its solid waste generation by 93 percent through a series of simple improvements in the way material was handled at the plant. These modifications included saving rinsewaters from tank and filter washings and converting from a one-stage to a two-stage rinsing of the vats in which chemicals were mixed. By using little water in the first rinse, concentrated chemical wastes could be collected for reuse. Such changes enabled Borden to close down a pollution control evaporation pond at the factory and to reduce its annual production of resin sludge from 350 cubic yards to 25 cubic yards, which alone saved Borden $48,750 annually in disposal costs.

Resources

The four resources listed at the end of Chapter 2 also are useful for Chapter 3.

Chapter 4

Creating Effective Source Reduction Programs

When you start to gather information about a plant in your community, you will realize that, as an outsider, you will probably never become familiar enough with the details of each production process to identify the most appropriate source reduction opportunities yourself. What you *can* do is promote the right atmosphere and procedures within the company that will enable the company to develop an effective source reduction program.

Company Atmosphere

A company that wants to explore source reduction opportunities has a goal of broadening its focus from managing to preventing pollution, thereby making its products more efficiently and reducing pollution in the community. To achieve this goal, companies must confront several ingrained habits and attitudes.

- A company must overcome a reluctance to interfere with processes that make commercially successful products. Inertia sets in when "the bottom line" shows a financially successful product.

- A company must stop considering pollution control costs as part of general company overhead and must assign them back to the production processes generating the pollutants requiring the control.

- A company must stop relying solely on separate pollution control departments for plant environmental protection. The specialty of such departments is pollution treatment and disposal, and their staff members typically have little control over the industrial processes creating the pollution in the first place and little contact with plant managers about the relationship between particular products and the pollution they create. (See box: The Limitation of Pollution Control Departments.)

- A company must shift the focus of its waste-related data collection from the end of the pipe to the plant processes themselves in order to reveal source reduction opportunities. The Office of Technology Assessment, INFORM, and others have found that ignorance of source reduction

---

The Limitation of Pollution Control Departments

The impact of dividing production and environmental protection responsibilities is perhaps best understood by picturing an enormous room full of industrial machines turning out thousands of widgets per hour. Many people are in the room, operating the machines, inspecting the widgets to see if they meet widget standards, and measuring their success by the number of widgets produced by a day's work. In the middle of the room is a large drain into which all the wasted oils, solvents, and other pollutants created by the process drain. At the end of the day, the drain is rinsed out and its contents disappear. There is a fat pipe outside the building that releases all the contents of the drain each day.

Environmental engineers at the end of the pipe measure the contaminants in the waste, consult appropriate regulations (to the extent they exist), and direct the flow of waste into treatment ponds, incinerators, landfills, and so on. These environmental engineers are not expected to determine how the waste products might be reduced or eliminated at the source, or even where the waste products come from. Their job is simply to clean up the pollution *after* it has been produced. This is the structural system that has focused efforts on end-of-the-pipe pollution controls at the expense of often inexpensive and easy source reduction strategies.

---

opportunities is fed by a culture within many plants (and environmental regulatory agencies) that does not reward the creativity and initiative of the very plant employees who are often best positioned to identify source reduction options — the employees involved in the day-to-day production processes.

A successful industry program requires changing the company's attitude about source reduction and promoting the flow of information about pollution problems, costs, and source reduction solutions. It comes down to developing an atmosphere in which information sharing and creative problem solving are emphasized and rewarded at all levels of the company.

Five Steps Companies Can Take

To create a plant atmosphere in which a source reduction program can flourish, plant managers can take five specific steps. They can adopt a waste management hierarchy, create leadership, develop incentives, collect waste-related data and include it in their cost accounting process, and develop plans. Encourage each company you study to take these five specific steps to improve information flow and to motivate employees to seek out source reduction opportunities.

Adopt a waste management hierarchy. The company should adopt the hierarchy discussed in Chapter 2 that places source reduction as the primary strategy, followed by recycling, treatment, and, as a last resort, disposal. Any company that is serious about source reduction should promote the primacy of source reduction in environmental protection through explicit policy. This policy should make clear that source reduction is the preferred, efficient option in all cases, not only for regulated pollutants or particularly costly inefficiencies. Such a policy will send a clear message to all employees — not just those with environmental responsibilities — that source reduction is the highest waste management priority.

Create leadership. The plant manager should assign oversight of the company source reduction program to a high-level employee who has the clout to get things done and whose job performance is judged by demonstrated source reduction achievements. This step is key to ensuring that the source reduction buck stops somewhere and the responsibility for running the factory at maximum efficiency with minimal materials loss does not float aimlessly around the company. The director of source reduction will also play a vital role in improving and coordinating the flow of information about source reduction throughout the company.

Develop incentives. A company should reward employees who excel at source reduction, just as it rewards those who develop new products or elevate production levels. Incentives — such as awards or bonuses — can be powerful tools for combating natural employee tendencies not to tamper with the status quo.

A company could also hold a contest to combine the processes of choosing source reduction projects and rewarding employee innovation. Employees would submit source reduction proposals for the contest and winners would be selected at a public ceremony. Dow Chemical, Inc., for example, an industrial leader in source reduction, held a contest in its Louisiana Division as part of its nationwide "Waste Reduction Always Pays" (WRAP) program. Twenty-three projects costing $3.3 million to implement were selected: within a year, they saved Dow $6.4 million, for a 193 percent return on investment.

Finally, companies should include source reduction efforts in job performance evaluations.

Collect waste-related data and incorporate it into the plant's cost accounting system. Before plant managers can develop effective source reduction plans for their facilities, they need to understand where wastes are being created and how much the wastes are costing the plant. They need three types of waste-related data to identify and prioritize source reduction opportunities and to track source reduction achievements: facility-wide data, process-specific information, and cost figures.

First, managers need *data on the movement and fate of chemicals used plantwide* in order to measure the plant's overall source reduction progress. That is, they need to know how much of each chemical enters the plant, how much is used in production, and how much is released as waste.

This plantwide information is obtained through a **facility-level materials accounting**: for each chemical raw material, managers compare the sum of the amount of the material already existing in plant inventory plus the material entering the plant with the sum of the amount of material consumed in plant processes plus the amount shipped out of the plant in products. The difference between these sums is the amount of the chemical that is released to the environment.

Second, a plant manager needs *information about the movement and fate of chemicals within each process* in order to identify specific sources of waste and activities leading to waste generation. This allows plant and process managers to determine what changes can be made to alter or eliminate these sources of waste release.

This information is obtained through a **process-level source reduction inventory** that includes, on a process-by-process basis, the same kind of comparison of material inputs and outputs that is done on a facility-wide basis, and a detailed analysis of how wastes are generated within each process.

To develop this process-level information base, the plant manager should select a high-level employee with a background in production operations, not pollution control, to direct annual source reduction inventories and develop plans for each production line in the company.

Since such a process has many steps, the director of the project will probably want to create a study team. In a large company, this team should include representatives from facility engineering, environmental engineering, safety and health, purchasing, materials and inventory control, finance, marketing, and product quality control. The team will:

- Collect background data on each production process, facility layout, waste stream generation, and pollution management costs.

- Conduct a plant survey to verify background data, identify additional waste streams, and collect data on actual operation and management practices.

- Sample waste streams to determine quantity and composition if this information is still unavailable.

- Conduct a quantitative assessment, called a **materials balance**, for each process to make sure that all major chemical releases to *all* environmental media are measured and accounted for. However, this procedure has limitations (see box next page: Materials Balance: Process and Limitations) and the process-level source reduction inventory must not rely solely on the quantitative materials balance of input and output. Relatively small levels of release can only be determined through actual measurement at possible sites of release and through asking the operations personnel a series of questions designed to determine the operator activities and status of equipment that might lead to release of materials in small quantities.

Third the manager must collect *data on the cost of materials and incorporate pollution-related costs into the plant's cost accounting system.* This involves assigning to each process the value of any "lost chemicals" revealed by the source reduction inventories, along with the cost of managing any pollution created by that specific process, thus helping managers understand that pollution adds to the costs of production.

Specifically, the cost of pollutants emitted by a process should be incorporated into the company's overall cost accounting system. Companies use cost accounting principles to measure the raw materials, capital, operating, and labor costs of a specific production line (see box page 17: Cost Accounting). These principles allow companies to balance the cost of production against sales of a given product and maximize profit. If pollution costs are factored into this cost accounting, then the plant manager will have a more accurate estimate of the true cost of the product. This procedure helps bring the environmental engineers "inside" the plant and brings home the importance of preventing waste generation to production managers and their employees, those best suited to identify source reduction

Materials Balance: Process and Limitations

A **materials balance** is a quantitative assessment of chemical inputs and outputs for individual processes. It is most useful for pinpointing relatively large releases of toxic and hazardous chemicals. Since a degree of inaccuracy is inherent in any measurement technique used, the materials balance is not as useful for identifying chemical releases in quantities that fall below this level of measurement accuracy.

The materials balance aims to account for every pound of a chemical that is (a) shipped to the process, (b) created or destroyed in the process, (c) delivered as a product from the process, and (d) wasted (whether it is an air, water, or solid waste). If the amount of wastes identified does not equal the difference between the amount of the chemical entering or being created in the process and the amount of the chemical leaving or being consumed in the process, then other sources of waste must exist.

For example, if a particular process receives 3 million pounds of a particular chemical as input, and if 2 million pounds of that chemical appear in the product or are consumed in the process, then the difference, or 1 million pounds, must be released as waste. If the plant manager was only aware of, say, 600,000 pounds of waste, it would be necessary to find other ways waste was being generated in the process.

It is important to know the level and nature of uncertainty in measurement techniques used for the materials balance for two reasons. First, for process-level source reduction inventories that seem to balance, the level of uncertainty may indicate that certain amounts of material may have been missed. On the other hand, if a such an exercise does not balance, but the difference is within the level of uncertainty, then it might be due to inherent measurement inaccuracies and not necessarily indicate a release into the environment.

For instance, in the above example, a facility inputs 3 million pounds of a chemical raw material for a particular process. Measurement techniques for such large quantities are typically accurate to within about 5 percent. In this case, that translates to an accuracy of +/– 150,000 pounds. Therefore, releases of this chemical from this process of up to 150,000 pounds would not be detected by the materials balance calculations.

opportunities. Like rewards, this helps ensure that each employee's "bottom line" reflects the overarching priority of reducing waste generation at the source.

Finally, the manager must *develop plans for source reduction activities in the plant.* Upon completion of the inventory and cost accounting, the manager should be able to list the waste streams in priority order for source reduction and identify applicable source reduction techniques. More information about the techniques described briefly in Chapter 3 is available from state technical assistance programs, from the EPA Pollution Prevention Information Clearinghouse, and from private consultants and chemical suppliers.

However, strategies for source reduction may be best developed by a company's source reduction team's own brainstorming. It is essential to share the results of each inventory with all of the operators of the production line in question. In many cases, they will not have had this kind of information before but they may be best equipped to use it effectively. These operators may have additional suggestions.

Cost Accounting	
Usually Includes	**Should Also Include**
Labor	Regulatory compliance
Fuel	Lost materials
Raw material inputs	Liability
Capital	Pollution control
Operating	and disposal

Ideas for making production more efficient should be selected based on their potential to prevent environmental and public health damage, their feasibility, and their costs. The selected projects should be organized into source reduction plans, forming a working agenda for source reduction in the company. They should contain goals and timetables for source reduction achievements, and their implementation should be closely monitored.

Resources

Dow Chemical Inc. (Department of Environmental Quality, Building 2030, Dow Chemical Inc., Midland, MI 48674. (515) 636-1000. Free.)

Dow has written several short descriptions of its source reduction program.

Institute for Environmental Studies. *A Handbook of Environmental Auditing Practices and Perspectives in North Carolina.* Chapel Hill, NC: University of North Carolina, 1985. (Available from Pollution Prevention Pays Program, North Carolina Department of Natural Resources, PO Box 27687, Raleigh NC 27611. (919) 733-7015.)

National Toxics Campaign. *The Citizens Toxics Protection Manual.* (Available from the National Toxics Campaign, 37 Temple Place, 4th floor, Boston, MA 02111. (617) 482-1477. Price: $35.00.)

In addition to contents described in resource list for Chapter 2, this book contains a useful discussion of important elements of an industry source reduction plan in Chapter 10.

North Carolina Pollution Prevention Pays Program. *Pollution Prevention Bibliography.* January 1986. (Available from Pollution Prevention Pays Program, North Carolina Department of Natural Resources, PO Box 27687, Raleigh NC 27611. (919) 733-7015.)

Lists 73 sources of information on environmental auditing.

Office of Technology Assessment. *Serious Reduction of Hazardous Waste.* (Summary available from the Office of Technology Assessment, United States Congress, Washington, DC 20510. (202) 224-8996. Free. Book available from the National Technical Information Service, 5285 Port Royal Road, Springfield, VA 22161. Price: $28.95 in paperback or $6.95 in microfiche, plus $3.00 per order for handling for either edition.)

The OTA report contains an extensive discussion of corporate source reduction.

United States Environmental Protection Agency. Pollution Prevention Information Clearinghouse. (Call (800) 424-9346.)

The Clearinghouse can provide information on source reduction techniques.

United States Environmental Protection Agency. *Waste Minimization Opportunity Assessment Manual.* Document EPA/625/7-88/003. Cincinnati, OH: Hazardous Waste Engineering Research Laboratory, July 1988. (Call (513) 569-7529.)

Chapter 5

Surveying Your Local Plant

This chapter will guide you step-by-step through the process of studying individual industrial plants to determine whether they are creating hazardous and toxic pollution and to assess whether they have made a commitment to source reduction. This process consists of four stages: considering which plants to study, conducting background research, interviewing plant representatives, and drawing conclusions. The worksheets in this chapter list the questions you will be asking in both the background research and plant interview phases; a complete set is also included at the end of this manual.

Considering Plants To Study

You may already know which plant or plants in your community you would like to assess. If not, there are a variety of criteria you could consider in selecting one to study. Since the data you will need to evaluate these criteria are not all available in the same place, the factors to consider are grouped below according to the information source you would consult. In every case, you would be looking for a plant that is significant to your community, whether because of its size, the amount of its releases, or some other factor.

Data available from the EPA's Toxics Release Inventory (see Appendix A for information on how to obtain these data):

- Proximity to residential areas
- Total amount of wastes released to a single environmental medium, or to all media
- Amount of wastes transferred off-site (a transportation safety concern)

Data available under the Resource Conservation and Recovery Act (RCRA) (call the RCRA Hotline, (800) 424-9346, for information on how to obtain these data):

- Facilities reporting little or no source reduction progress on their Waste Minimization Reports available from EPA (see Appendix E)
- Size of on-site hazardous waste management facilities with RCRA permits
- Number of RCRA hazardous waste manifests submitted (a transportation safety concern)
- Recent submitters of permit applications to construct an on-site hazardous waste management facility

Data available from state environmental air and water permit offices:

- Number of air and/or water discharge permits
- New permit applications
- Number of permit compliance violations

Other data available in government offices:

- Number of Occupational Safety and Health Administration (OSHA) violations (state)
- Applications to construct a new manufacturing facility or expand an existing one (state)
- Number of Premanufacturers Notifications (PMN) submitted by companies under the Toxic Substances Control Act (TSCA) for the manufacture of new products or processes (federal)

Conducting Background Research

Your discussions with company representatives will be most effective if you have learned as much about the plant and its activities as you can before you actually meet. Background research allows you to develop an overall picture of the plant you are studying, its compliance status with pollution control regulations, and the pollution it is generating.

Profile of the plant (Worksheet 1)

You will first want to gather general information about the characteristics of the plant you are studying. This information is important for three reasons. First, data about production, employment, and sales volume can help put the company's program in the right perspective. It may be much more impressive for a small company to eliminate 100 tons of a given contaminant than for a larger one to achieve the same result. Second, obtaining such information could prove invaluable in comparing the performance of one plant with others that share the same characteristics. Finally, it is important to determine who actually owns the plant in order to identify who is responsible for the company's environmental policy.

Sources:

Company annual reports (available from the company)

State manufacturing directories (available from state Department of Commerce)

Dun & Bradstreet reports (financial reports on individual companies available to subscribers; you can get information from local Dun & Bradstreet office in nearest large city)

Securities and Exchange Commission "10k" reports (available from local SEC offices or from SEC headquarters in Washington: 450 5th Street NW, Washington DC 20549)

Federal Toxics Release Inventory (see Appendix A)

(Note: Some data may not be available through public information sources and can only be obtained during the plant interview.)

Plant permit compliance and applications for new permits (Worksheet 2)

It is important to obtain information about the extent to which the plant has met the requirements of the various permits controlling its management or release of toxic waste to the air, land, or water. This information can be useful in several ways.

- First, a major violation or pattern of violations can be helpful in convincing the state and the local community to persuade the plant to adopt a comprehensive source reduction program.

- Second, you may be able to convince the plant manager that compliance problems can best be avoided through a program that reduces the quantity of chemical releases that must be managed under complex permit conditions.

- Third, circumstances are often most favorable for persuading a plant to adopt a source reduction plan when it is seeking approval from the state for a new permit or for some change to its existing permit conditions. Most states have procedures that provide for a public hearing or at least an opportunity for public comment on a company's application for a new or modified permit.

Remember, however, that source reduction should extend to all chemical releases, regardless of whether they are subject to permit requirements. For example, many chemicals in the Toxics Release Inventory are only regulated when discharged into one environmental medium or are not regulated at all.

Worksheet 2, including the table on specific violations, can be completed by reviewing permit files at the state environmental office or by speaking directly to the officials in that office who write the permits. An example of how to fill out the table is also included.

Sources: The people in your state government who actually wrote the pollutant discharge permits for the plant you are studying are one of the best sources of information you have. If you need help identifying the appropriate contact in your state agency for permit information about specific facilities, you can contact professional organizations.

Solid Waste	Association of State and Territorial Solid Waste Management Officials 444 N. Capitol Street, NW Suite 388 Washington, DC 20001 Tel: (202) 624-5828 Attn: Julie Graebe
Water	Association of State and Interstate Water Pollution Control Administrators 444 N. Capitol Street, NW Suite 330 Washington, DC 20001 Tel: (202) 624-7782

Air   Waste Management Association
                       PO Box 2861
                       Pittsburgh, PA 15230
                       Tel: (412) 232-3444

                       State Air Pollution Control Administrators
                       444 N. Capitol Street, NW
                       Suite 306
                       Washington, DC 20001
                       Tel: (202) 624-7864
                       Attn: Nancy Kruger

Individuals at these organizations should be able to provide you with a starting place in locating the state office that maintains information about permits for air and wastewater discharges and disposal on land. If you plan to be involved in surveys in more than one state, you may wish to purchase the membership directories that each of the above organizations produces. However, in some cases, these directories list only the administrator or main office for the program and may not provide the name and address of the particular individual responsible for information about specific facility permits.

Plant pollution profile (Worksheet 3)

Before you meet with company officials, you will want to know what hazardous and toxic materials the plant is releasing, and where these releases are going. Worksheet 3 is designed to help you organize and understand this information. In it, you will list the amounts of each pollutant released to the different segments of the environment.

Much of the data required for completion of the table is available on the Toxics Release Inventory (TRI) forms companies must fill out for individual toxic or hazardous chemicals. (See Appendix A for an explanation of the TRI process and a sample of form "R," the report form.) Indeed, as long as the company you are interested in is a *manufacturer* (not an incinerator, local gas station, or some other nonmanufacturing operation) and meets TRI reporting requirements (such as size and amounts of chemicals used), the TRI form is your best source of information. This form, supplemented by a helpful state permit writer (see the source discussion in the section on permit compliance) may be all you need to develop adequate background on a plant.

You will probably not be able to find all the information requested in Worksheet 3 in publicly available sources and will therefore have to ask company representatives for it when you meet with them. Specifically, the information about quantities of waste sent off-site for recycling is not required on the TRI form. In addition, you will need to ask company representatives for data on chemicals that are not covered by TRI or are released in quantities below the TRI threshold. Also ask for information on waste streams not reported under the Resource Conservation and Recovery Act

In Worksheet 3, the first line should show the current amount of all hazardous and toxic wastes actually generated during production, *before* any kind of out-of-process recycling, treatment, or disposal. Industries are given the option of reporting such data on individual chemicals on the TRI, but are not required to do so. If these numbers are *not* available on the TRI form, ask company representatives for the data in your interview. In many cases, they may refuse to provide such information, or may not have measured it. Nevertheless, this information is important since a source reduction program should be specifically designed to reduce that number. (Remember that source reduction only covers those reductions that occur prior to recycling, treatment, or disposal.)

You may also want to ask company representatives for graphs of overall emission trends for land, water, and air pollution, or for specific chemicals of concern. This information enables you to study the *patterns* of pollution at the plant over a number of years. It will *not* provide a measure of source reduction, however, since emissions reductions may be achieved through treatment and recycling, as well as source reduction.

Sources:

> Permit applications and other publicly accessible files (available at state or regional environmental agency offices)

> TRI forms (for availability, see Appendix A)

> Material Safety Data Sheets and Chemical Inventories (available from emergency planning groups; see Appendix D)

> Company representatives

> RCRA annual Hazardous Waste Generator Reports and RCRA Biennial Waste Minimization Forms (see Appendix E for availability of RCRA information)

> Discharge Monitoring Reports (available from state permitting offices; for wastewater discharges to a **publicly owned treatment work** (a **POTW**, or public sewage facility), available from the POTW itself)

(Worksheets 1 through 3 on following pages)

A. Identification

    i.   Plant name _____

    ii. Plant ownership or affiliation (include most recent changes in ownership) _____

       _____

B. Output and employment

    i.   Annual sales data (most recent year) _____

    ii. Annual production volume (most recent year)* _____

    iii. Number of plant employees in (a) administration _____

       and (b) production _____

    iv. Date operations began_____

    v. Major changes in output and type of product in the past ten years _____

       _____

       _____

C. Product type

    i.   Categories of products manufactured at the plant (include Standard Industrial Code (SIC) for product type; these codes are listed in Table A-1 in Appendix A)

       _____

       _____

       _____

    ii. Type of production process used: continuous or batch (**Continuous** processes are dedicated to continuous production of a single product; **batch** processes produce different products at different times. Continuous operations usually generate less waste per pound of product.)*

       _____

    iii. Changes in production output planned that might proportionally affect the quantity of toxic releases*

       _____

       _____

*May not be available in public databases; this information may have to be obtained during plant visit.

<table>
<tr><td>

Plant Background Information

Plant: _____

Date Filled Out: _____

</td><td>

</td></tr>
</table>

A. Environmental Non-Compliance Profile

Wastewater Discharge (specify pollutants)

Pollutant	Description of Violation

Air Emission (specify pollutants)

Pollutant	Description of Violation

Hazardous Waste Treatment, Storage, Disposal Facility (TSD) Permit *

Facility Type/ Capacity	Description of Violation	Operating/ Not Operating

*Also required for hazardous waste recycling facilities

B. Violations

   i.   Circumstances in which the plant has exceeded limits established in its permit.

   ii.   If the violation was sudden or accidental, what steps has the plant taken to prevent a recurrence?

   iii.   Have any permit violations led the plant to take a closer look at source reduction?

C. Permit Applications

   i.   Does the plant have any applications pending before state or local government for modifications to any existing permits? If so, what are they?

   ii.   Does the plant have any applications pending before the state or local government for a new permit (e.g., construction of a new incinerator or landfill)? If so, what are they?

   iii.   In reviewing the plant's application for a permit modification or new permit, has the state inquired whether a source reduction program exists, or required one? If so, explain.

   iv.   Has the plant submitted a premanufacturer's notification (PMN) for any new products or processes as required by the Toxic Substance Control Act (TSCA)? (Requests for TSCA related information should be sent to EPA headquarters, ATTN: Jeralene Green (A-101).)

Plant Background Information

Plant: _____Fossil-To-Fabric, Inc._____

Date Filled Out: _____March 17, 1990_____

A. Environmental Non-Compliance Profile

Wastewater Discharge (specify pollutants)

Pollutant	Description of Violation
Phenol	Exceeded discharge limit of 9.06 kg/day for 24 of the last 52 weeks.
Chromium	Failed to report discharge information on monthly monitoring reports.

Air Emission (specify pollutants)

Pollutant	Description of Violation
Methylene Chloride	Exceeded number of hours allowed to operate emission source.
Xylene	This air pollutant not found in the facility's current permit.

Hazardous Waste Treatment, Storage, Disposal Facility (TSD) Permit *

Facility Type/ Capacity	Description of Violation	Operating/ Not Operating
Wastewater Treatment/ 1 million gallons/day	Failed to submit groundwater monitoring report	Operating
Storage	Leaking containers	Operating

*Also required for hazardous waste recycling facilities

27

Worksheet 3

Plant Pollution Profile

(All amounts in pounds per year)

Plant Background Information

Plant: _____

Date Filled Out: _____

Pollutant Name/CAS Number					
Total Releases	Before end-of-pipe management				
	After end-of-pipe management				
Air	Fugitive emissions				
	Stack emissions				
	Total Released to Air				
Water	Released to surface water				
	Underground injection				
	Discharge to Publicly Owned Treatment Works (POTW)				
	Total Released to Water				
Land	On-site disposal				
	Off-site disposal				
	Total Released to Land				
Recycling and Treatment	On-site Recycling				
	Off-site Recycling				
	On-site Treatment				
	Off-site Treatment				
Source of Data					

Plant Pollution Profile

(All amounts in pounds per year)

Plant Background Information

Plant: ___Fossil-to-Fabric, Inc.___

Date Filled Out: ___March 17, 1990___

Pollutant Name/CAS Number		Ethylbenzene 100-41-4	Dodecyl Phenol 27193-86-8	Hazardous Waste Solids Not Otherwise Specified
Total Releases	Before end-of-pipe management	?	?	?
	After end-of-pipe management	>120,000	636	210,080
Air	Fugitive emissions	1-499	?	?
	Stack emissions	1-499	636	?
	Total Released to Air	2-998	636	?
Water	Released to surface water	0	?	?
	Underground injection	0	?	?
	Discharge to Publicly Owned Treatment Works (POTW)	2-998	?	?
	Total Released to Water	2-998	?	?
Land	On-site disposal	0	?	?
	Off-site disposal	0	?	?
	Total Released to Land	0	?	?
Recycling and Treatment	On-site Recycling	0	?	?
	Off-site Recycling	?	?	?
	On-site Treatment	0	?	?
	Off-site Treatment	120,000	?	210,080
Source of Data		1987 TRI data	1986 State Air Emissions Report	1986 RCRA Generator Annual Report

Preparing for the Plant Interview

Once you have gathered background data, you are ready to contact plant officials and seek an interview. Since public records do not provide a detailed compendium on source reduction activities at individual facilities, an open dialogue with company officials, including the plant manager, is necessary. Further, by creating a positive relationship with company officials, you will be in a better position to encourage them to focus on source reduction strategies. This section guides you through the steps of setting up a meeting and breaking the ice; it also alerts you to arguments company officials may use.

Setting up a meeting

Now you are ready to set up a meeting with the plant manager. Figure 5-1 summarizes the steps to follow in trying to set up that meeting and in responding to a company refusal to meet with you. Since your goal is to establish a cooperative relationship with company officials, it is important that your initial communications with the company, both by phone and by letter, be professional and positive. A sample letter to a plant manager is shown on the next page. It is useful to include a list of sample questions so the plant manager can prepare for the meeting and designate the appropriate company employees to meet with your group.

While you are setting up the meeting, you can also be putting your interview team together. It should include a respected member of the community and a person with technical knowledge. A team of about two or three people is usually an effective size.

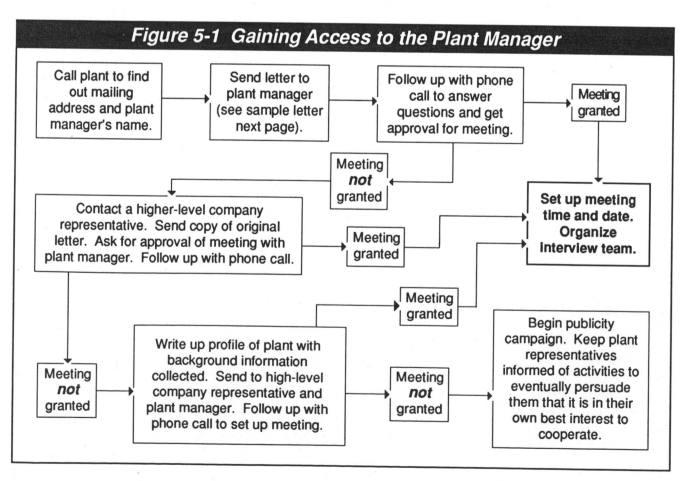

Figure 5-1 Gaining Access to the Plant Manager

Sample Letter to Plant Manager

(Group Name)
(Address)
(Phone Number)
(Date)

(Plant Manager's Name)
(Title)
(Plant Name)
(Address)

Dear *(Plant Manager's Name)*:

(Group Name) has begun a program to learn about what generators of hazardous and toxic waste in our region are doing to reduce at the source the amounts of the wastes they release to air, land and water. We will be educating interested citizens and others about the extent of such source reduction activity. We invite you to meet with us to discuss what source reduction measures have been taken or are planned at *(Name of Plant)*.

Source reduction, according to the congressional Office of Technology Assessment (OTA), covers "in-plant practices that reduce, avoid, or eliminate the generation of hazardous waste so as to reduce risks to health and environment." OTA includes recycling that is "an integral part of the ... industrial process," or "in-process" recycling, in its definition. It does *not* include recycling that involves moving wastes to another area of the plant or to an off-site location.

Numerous case studies by government, private, and industry groups have shown that such source reduction measures often save companies money.

We are not concerned with proprietary information but with the overall picture of source reduction programs and achievements at your facility. The enclosed list will give you an idea of the kinds of questions we will be asking. It should help you prepare for our meeting and designate the appropriate plant officials to meet with us.

In case you are not familiar with *(Group Name)*, we are a group of citizens from the *(name of area)* area who work together to improve the environmental quality of our region. Our previous activities have included *(some recent activities)*.

We hope that with your cooperation and that of other hazardous waste generators in the region, we can move toward reducing the region's overall waste burden.

We look forward to hearing from you soon.

Sincerely,

Enclosure for Sample Letter to Plant Manager

Sample Questions for Plant Interview

1. **Responsibility for source reduction policy**

   o   Does your company/plant have an official, written policy on hazardous and toxic waste management and does it include a hierarchy of management options?

   o   Where does source reduction stand in this hierarchy?

   o   Which division/office/person oversees the implementation of this policy?

2. **Corporate mechanisms to implement source reduction**

   o   What are the elements of your plant's source reduction program? For example, is there any source reduction training and incentives program for process operators?

   o   Does the plant collect waste generation data at the process level? What is the scope of this information collection?

   o   Is there a full cost accounting system for hazardous and toxic wastes?

   o   Who, within the plant, is responsible for reducing the amount of hazardous and toxic waste generated?

   o   What actions have been taken to reduce the hazardous and toxic waste generated by the plant? For each action, what wastes were reduced, how was the source reduction opportunity identified, what factors influenced the choice of source reduction techniques used, and how much waste reduction was achieved?

Creating a positive relationship

Probably the most effective tool for getting the information you need from company officials on significant source reduction activities will be the development of a friendly, cooperative relationship with them. The following pointers can help you establish a positive relationship.

- Do the background research thoroughly and carefully before setting up an appointment with company representatives. Let them know that you have tried to gather as much information as possible prior to taking their time with a survey.

- Carefully phrase questions about compliance with current environmental laws and permits. Aggressive questioning in this area may send the conversation in a defensive direction, and you will not get the information on source reduction that you want.

- Make clear that you are not interested in a company's trade secret information. You are trying to get a basic source reduction picture, rather than technical details on individual processes.

Of course, if the company attitude toward citizen inquiry prevents the creation of a positive relationship, you might want to consider other tactics.

Arguments to beware of

As you conduct your interview, company representatives may try to explain why the company does not have an adequate source reduction program. Several arguments are frequently offered; with enough background information, you will be able to avoid being put off by them.

"We're in compliance with the law." Companies traditionally evaluate their environmental management program according to whether it is successful in preventing permit violations or enforcement actions brought by regulators or citizen groups. Many companies have designed audits to assure compliance with laws and regulations, and the person you are interviewing may attempt to share this information with you. These compliance audits are a welcome sign that companies are taking their legal responsibilities seriously. However, they are *not* a substitute for a source reduction program because existing regulations are focused on end-of-the-pipe control strategies and affect only a fraction of the pollutants generated by industry.

- Many chemical substances remain unregulated. For example, the EPA has issued regulations for only seven out of the thousands of possibly toxic air pollutants.

- Many sources are unregulated. For example, the Clean Air Act and the Clean Water Act regulate only point sources of contamination (specific planned emission points such as smokestacks and discharge pipes). They do *not* capture the large volume of contaminants released as fugitive air emissions (from unintentional release points such as leaky valves or evaporation tanks), or as runoff from the plant's grounds.

- Finally, even the existing regulations allow a relatively large volume of pollutants to be released to the environment.

In general, a company's source reduction program should *not* be limited by whatever minimum is needed to comply with the existing laws. You might want to ask the company representatives if they believe that further regulations are needed to get the industry to make additional reductions in the generation of toxic and hazardous pollutants (and if so, what kind). If not, then the industry will have to clearly demonstrate — as you are asking them to do — how it is moving *beyond* the limits of existing regulations.

"The pollutants released from this plant pose a very low risk to local citizens." You are likely to hear that the plant's toxic releases pose a relatively low risk to the surrounding environment. The company representative may point out that the plant is already required to comply with occupational exposure limits to protect human health (sometimes called "threshold limit values") and that exposures to inhabitants of the surrounding neighborhood are many times lower than these occupational limits. While it is essential that companies protect the health of their workers, doing so may not be sufficient to protect neighboring communities or the environment.

- Plant workers are often young healthy males with a relatively high tolerance for exposure to contaminants. Usually they are only exposed eight hours a day, five days a week. The population at large, in contrast, includes pregnant women, babies, the elderly, and other groups with relatively low chemical tolerances. These people are often exposed continually and involuntarily to contaminants.

- Much of the data about the health effects of exposure to chemical contaminants is speculative. For example, exposure limits may be based on animal studies and limited by unprovable assumptions about the relationship between the dose to which an individual is exposed and the likely health effects. These studies also are usually limited to individual chemicals and do not consider the multiplier effects that may occur when several chemicals interact in the environment. Further, the threshold limit values developed by the government are based on workday, not round-the-clock, exposures.

- Concern about chemical contaminants should not be limited to human health. In many cases, while the health risks may be relatively low, more serious side effects may be felt by the natural environment, through ecological damage to rivers, lakes, and soils, for example. The Great Lakes provide a classic example of an ecosystem where contamination by many different chemicals from many different sources has produced serious damage.

Given these risks and uncertainties, and the potential economic benefits of source reduction, the wiser course is to take preventive actions now that might prevent problems from arising in the future.

"Source reduction at my plant is technically impossible, or too expensive." As discussed in Chapter 3, many of the opportunities for cutting the generation of potentially toxic wastes involve no-technology or low-technology options, such as substituting raw materials or improving maintenance procedures. Evaluations by INFORM, OTA, and private industries have shown that *some* source reduction opportunities may be found in almost every case where a thorough process-level source reduction inventory has been undertaken. Accordingly, if a company representative cites economic or technical barriers, you will want to ask questions about how this was determined.

- Have you conducted a thorough process-level source reduction inventory for each process used at this facility?

- If you encountered cost barriers, do you use cost accounting principles that reflect the costs that pollution adds to a particular process?

- If you encountered technical barriers, have you sought technical assistance from the state, local government, or your trade association? If so, please describe the source of technical assistance and the type of assistance received (for example, information packet sent by mail, on-site technical assistance, referral to equipment vendor)?

- In your evaluation of source reduction options, did you consider no-technology or low-technology choices as well as more significant changes to production or design? If so, please describe.

"I can't share that with you because it's a trade secret." Companies are often reluctant to share the details of a specific technological process out of concern that it may provide some advantage to a competitor. While this concern is often well-founded, at times it is the result of a general tendency to "play it safe" by throwing the "trade secret blanket" over large quantities of relatively harmless information. If you are faced with this claim, make the following points:

- You are *not* asking for a detailed description of a "proprietary" process. The company accomplishments chart (Worksheet 6) asks only for a general, categorical description of the type of process used and the source reduction savings that resulted.

- Many companies have already provided this kind of data in response to a survey sent out by Tufts University's Center for Environmental Management. Examples of companies that responded to the survey are Dow, Monsanto, General Electric, and Union Carbide.

- More than 2,000 industrial facilities have already made such information available by filling out the optional source reduction form attached to their 1987 Toxics Release Inventory form.

- Finally, other companies have provided much more detailed information about specific source reduction processes and the results they achieved. This information has been compiled by North Carolina's Pollution Prevention Pays Program.

"We have already achieved all the source reduction possible." Research by OTA, INFORM, and others has concluded that most plants have not systematically looked for opportunities to reduce waste generation. Does this company have a systematic approach in place? If the company has not undertaken any of the organizational strategies discussed in Chapter 4, the chance that it has reached its source reduction potential is close to zero. Ask the company representative what success the company has achieved and how it seeks new opportunities.

"We measure source reduction by waste streams, not specific chemicals." The Resource Conservation and Recovery Act (RCRA) regulates solid hazardous waste as "waste streams" made up of a number

of chemical components. For example, RCRA K104 wastes are "combined wastewater streams generated from nitrobenzene/aniline production."

However, source reduction opportunities and achievements are most effectively evaluated by conducting a chemical materials balance (see Chapter 4) for each process in a plant. The best way to conduct such a materials balance is on an individual chemical basis because complex mixtures can be altered substantially as they pass through the many possible steps in a process and thus become difficult to trace. For example, if the more volatile portions of a complex solvent mixture evaporate in a process, two completely different streams are created (the volatile compounds in the air and the residual liquids). Thus, materials balance measurements become overly difficult if specific component chemicals are not the units being measured.

Of course, you will have to accept waste stream data if it is all the company has, but be sure to ask the company representative about the usefulness of this kind of information in measuring source reduction.

- Are you measuring air pollution and certain types of wastewater discharges that are not released as a "waste stream," or is your source reduction program only geared toward solid wastes regulated under RCRA? (If so, it is not a truly multimedia source reduction program.)

- How do you know whether you have reduced the toxic components of a given waste stream that are the cause for concern? Or have you simply "dewatered" the waste stream (that is, reduced its volume by removing water) or taken out some other harmless component?

- If the company is already reporting **chemical-specific** emissions for TRI (emissions for individual chemicals), why is it not possible to measure source reduction on a chemical-specific basis?

Interviewing Plant Representatives

Your interview with plant representatives is designed to help you learn about the company's source reduction activities. You will notice, however, that many of the questions on the plant visit worksheets focus on institutional structures that encourage source reduction, not technical assessments of the company's source reduction program. As discussed in the introduction, unless you have an extensive technical background, the most effective role you can play is to open a dialogue on the institutional questions. If you achieve success on these, you will have the whole company seeking out individual source reduction options. Of course, you will also need to hold the company accountable for overall achievements.

Measuring the corporate commitment to source reduction (Worksheet 4)

The strategies for achieving source reduction range from simple housekeeping measures to much more sophisticated changes in product or process design. Using these strategies effectively requires fundamental changes in the corporate culture, bringing environmental engineers, production managers, and other employees together to make the industrial process less "waste-intensive." Without a high-level and continuing commitment to source reduction, any successes claimed may be temporary or isolated events.

It is also important when discussing source reduction with corporate officials to clarify early on what activities they understand source reduction to encompass. The box in Chapter 2 on Why "Source Reduction" Is Not Necessarily "Waste Reduction" or "Pollution Prevention" explains how source reduction differs from other waste management practices. Earlier surveys have shown that many companies have understood source reduction to include almost any activity except land disposal (incineration, for example).

Waste-related data collection (Worksheet 5)

A systematic program for monitoring the amount of toxic chemicals released to the environment is critical to a source reduction program because it provides the information needed to identify source reduction opportunities and to measure source reduction achievements over time. Without a system for measuring how much waste generation has been prevented, it is impossible to determine the effectiveness of specific source reduction methods.

Thus, the existence of a systematic data-gathering program in a company or plant reflects a serious commitment to long-term source reduction. The questions on Worksheet 5 will help you assess the extent to which the plant you are studying has instituted key elements of a waste-related data collection program: facility-level materials accounting, process-level source reduction inventories, and cost accounting. These elements are explained in Chapter 4.

Measuring company accomplishments (Worksheet 6)

After finding out about the plant's institutional structure and its program for collecting waste-related data, both of which give you information about the company's commitment to source reduction, you still will want to measure the company's actual source reduction accomplishments. Ask the company representative to help you fill out the achievements profile (Worksheet 6) for each pollutant that has been reduced or eliminated through source reduction. The worksheet can also be used to gather information about the company's future source reduction plans. (An example of how to fill out the worksheet is shown.)

Corporate incentives and barriers to source reduction (Worksheet 7 — optional)

Information about the factors that motivate a plant to adopt an effective source reduction program can be particularly important in helping your state or local government to identify barriers to source reduction. Worksheet 7 covers *why* a plant might be achieving source reduction, or why it is not, rather than *whether* it is achieving source reduction. You may consider these questions optional because they are not essential for your activities at the local level. However, they provide you with additional information about motivation that will be useful if you plan to pursue governmental options such as those in Chapter 6.

(Worksheets 4 through 7 on following pages.)

Plant Visit Information

Plant: _____

Date Of Visit: _____

Company Representatives: _____

Question	Company Response
A. Source Reduction Policy	
i. Does your company have a working definition of source reduction? What is it? When was it first established?	
ii. Does your company have a written policy favoring source reduction as the most desirable waste management option? If so, may we have a copy? Who establishes the policy?	
iii. Is the policy corporation-wide or plant-specific?	
iv. Is your source reduction program a multimedia program? That is, does it cover chemical releases to air, water, and land?	
v. Do you have specific source reduction plans for specific chemicals?	
vi. For specific industrial processes?	
vii. Does your plan include specific source reduction goals? How are they measured?	
viii. Are source reduction options considered during the engineering phase when planning new product lines?	

Question	Company Response
B. Company Leadership	
i. Is there an individual responsible for source reduction within your plant? If so, what is that individual's title and technical/program management background?	
ii. Is there a department or division at the company or plant that is responsible for source reduction? If so, how is it staffed, where is it located, and what are its general responsibilities?	
iii. How do this division and individual fit into the overall corporate structure? That is, to whom do they report and who reports to them?	
iv. To what extent are the managers and operators of your production processes involved in source reduction? For example, do they identify opportunities, implement changes, or measure progress? (Research shows that many source reduction opportunities are identified by production personnel.)	
C. Source Reduction Incentives	
i. Is there an incentive system that encourages employees to come up with suggestions for source reduction? What are the incentives?	
ii. Is there a source reduction training program? If so, please describe.	
iii. Is source reduction information transmitted throughout the plant/company in the form of newsletters/staff meetings? May we see written examples?	

Plant Visit Information

Plant: _____

Date Of Visit: _____

Company Representatives: _____

Question	Facility Level (facility-level materials accounting)	Process Level (process-level source reduction inventory)
A. Facility and Process Level Data		
i. Does the company collect waste-related data at this plant?		
ii. How are materials tracked (as specific chemicals, by waste categories, as products, etc.)?		
iii. Which chemicals and/or waste categories are tracked?		
iv. Does the company track materials released to all environmental media (air, land, and water)?		
v. Are materials tracked in each environmental medium regardless of whether they are regulated in them?		
vi. Are materials tracked in both production and non-production areas of the plant? (Nonproduction areas include storage, loading, transfer, and pollution control sites.)		
vii. Are chemical-specific materials balances performed to assure identification of all major sources and quantities of waste generated?		

Question	Facility Level (facility-level materials accounting)	Process Level (process-level source reduction inventory)
viii. If a materials balance is not performed, why not?		
ix. Are the nature and levels of uncertainty in the measurement methods used identified?		
x. How are these levels of uncertainty factored into the analysis of the overall results? (Make sure you get some numbers on the amount of material not captured by the measurement methods used.)		
xi. Who sees the results of the data collection program?		
xii. What are the data used for?		
xiii. How do these data factor into evaluations of both the source reduction policy and the specific actions taken to achieve reductions in waste generation?		
xiv. How frequently are the data collected?		
xv. Who performs the collection procedure? What is this person's educational and professional background?		

Question	Facility Level (facility-level materials accounting)	Process Level (process-level source reduction inventory)
xvi. Have you computerized this information?		
xvii. Have you established a base year from which present and future accomplishments are measured? If so, what is that year?		

Question	Company Response
B. Progress Reports	
i. Does the plant prepare progress reports?	
ii. How do you measure source reduction accomplishments (e.g., by pounds per unit of production)?	
C. Full Cost Accounting	
i. Are waste-related costs accounted back to their source in the company's cost accounting system? Or does the company treat environmental costs as a fixed overhead expense?	
ii. Are wastes released to all environmental media included in the cost accounting?	
iii. Are costs designated by specific chemicals, chemical categories, or waste categories?	

Question	Company Response
iv. What types of costs are captured by the full cost accounting method you use: a. Materials (e.g., the costs of wasted starting materials and lost products)?	
b. Waste handling (e.g., capital and operational expenses for on-site recycling, treatment, storage, or disposal facilities; transportation and other expenses for wastes sent off-site)?	
c. Regulatory compliance?	
d. Insurance?	
e. Future liabilities from waste generation (e.g., accidents, worker illness, or waste site clean-up)?	
f. Public and customer relations dealing with waste issues?	
g. Other (please specify)?	

Plant Visit Information

Plant: _____

Date Of Visit: _____

Company Representatives: _____

Achieved or Planned			
Specific Waste Reduced (hazardous or nonhazardous)			
Waste Medium			
Source Reduction Type (see code below)*			
Source Reduction Practice			
Percent Waste Reduced			
Amount Waste Reduced			
Motivation			
How Identified			
Comments			
O P T I O N A L	Change in Yield			
	Dollars Saved			
	Dollars Spent			

*Source reduction types: CS, chemical substitution; EQ, equipment change; OP, operational change; PR, product reformulation; PS, process change.

Plant Visit Information

Plant: Fossil-to-Fabric

Date Of Visit: May 12, 1990

Company Representatives:

Achieved or Planned	Achieved	Planned	
Specific Waste Reduced (hazardous or nonhazardous)	Formaldehyde (hazardous)	Acrylonitrile (hazardous)	
Waste Medium	Water	Air	
Source Reduction Type (see code below)*	OP, EQ	EQ	
Source Reduction Practice	Recirculate formaldehyde-contaminated vacuum pump seal water back to process. Carbon steel pumps replaced with stainless steel pumps that do not corrode as concentration of formaldehyde increases.	Install condensers on ABS plastic production unit.	
Percent Waste Reduced	98%	50% (projected)	
Amount Waste Reduced	6,000 lbs/year	100,000 lbs/year (proj.)	
Motivation	Local POTW imposed a limit of 50 ppm	To comply with good neighbor agreement with local environmental group	
How Identified	Employee suggestion	Plantwide program to identify fugitive air emissions	
Comments		Timetable 9/90	
OPTIONAL — Change in Yield	0.005%	?	
OPTIONAL — Dollars Saved	$420/year	$75,000/year (projected)	
OPTIONAL — Dollars Spent	$20,000	$10,000 (projected)	

*Source reduction types: CS, chemical substitution; EQ, equipment change; OP, operational change; PR, product reformulation; PS, process change.

Plant Visit Information

Plant: _____

Date Of Visit: _____

Company Representatives:_____

A. Establishing Priorities

Please indicate the order of importance of each of the following factors in terms of its influence on your source reduction program.

☐	Reducing releases of the most hazardous chemicals
☐	Complying with regulations
☐	Avoiding potential future liability
☐	Saving raw material costs or otherwise improving process efficiency
☐	Avoiding waste management costs
☐	Other (please specify)

B. Incentives for Source Reduction

Please identify which incentives influenced your plant's decision to implement source reduction, and indicate their relative importance on a scale of 1-10 (10 being most important).

i. The Toxics Release Inventory required by EPA, and the fact that it will be made public, create a strong incentive for companies to implement source reduction policies.

Yes ☐   No ☐   Importance (1-10) ☐

ii. Source reduction makes economic sense because it can save raw materials and help to make production more cost-effective.

Yes ☐   No ☐   Importance (1-10) ☐

iii. The state has provided us with information or technical assistance that helped us to reduce waste at the source.

Yes ☐   No ☐   Importance (1-10) ☐

iv. Trade associations or other industry sources provided us with information or technical assistance that helped us to reduce waste at the source.

Yes ☐   No ☐   Importance (1-10) ☐

v. Increasingly stringent regulations have made it necessary to find innovative ways to reduce the cost of compliance.

Yes [ ]   No [ ]   Importance (1-10) [ ]

vi. The state (or local government) has persuaded us to put a source reduction program into place as a result of an enforcement action or negotiation over permit conditions.

Yes [ ]   No [ ]   Importance (1-10) [ ]

vii. We look for ways to reduce pollutants that are not currently regulated to cut down on the need for future regulation or to avoid the need for compliance costs if such materials are regulated in the future.

Yes [ ]   No [ ]   Importance (1-10) [ ]

C. Barriers to Source Reduction

Please identify those statements with which you agree and indicate their relative importance on a scale of 1-10.

i. We have experienced significant engineering or technical barriers to source reduction.

Yes [ ]   No [ ]   Importance (1-10) [ ]

ii. We have experienced cost barriers to source reduction (please specify).

Yes [ ]   No [ ]   Importance (1-10) [ ]   [ ]

iii. We have identified serious regulatory obstacles to source reduction (please specify).

Yes [ ]   No [ ]   Importance (1-10) [ ]   [ ]

iv. We could benefit from state technical assistance.

Yes [ ]   No [ ]   Importance (1-10) [ ]

Drawing Conclusions and Using Them

Once you have gathered background information from all available sources and have learned as much as possible from your meeting with company representatives, you will be able to draw conclusions about the plant you have studied. You will know whether it:

- is creating toxic or hazardous pollution;

- is in compliance with existing environmental laws;

- has made a policy and staff commitment to reducing waste at the source;

- has established a systematic program to identify source reduction opportunities and associated cost savings;

- has assessed source reduction progress; and

- has outlined source reduction plans for the future.

What happens next? You will want to use the data you have gathered to encourage plants to adopt source reduction practices. While the primary purpose of this guide is to help you obtain this information, a summary of further activities you could undertake follows. For more details and practical suggestions, you may find it helpful to look at guidebooks describing strategies for citizen organizing published by the Citizens Clearinghouse for Hazardous Waste, the National Toxics Campaign, and Greenpeace (see Resources.)

Whether you assess the study plant to be a shining star or a profligate wastrel, publicizing the study findings through local news media or other outreach campaigns might prove effective in promoting source reduction in your area. If the study plant has established a successful systematic source reduction program, public recognition might set this plant as an example for others to follow. On the other hand, if the study plant has made few efforts to identify and pursue source reduction opportunities, a clearly presented comparison of the steps outlined in this guide, and the steps lacking at the study plant, could spur the company, and others like it, to take more aggressive action.

Your findings might also be the basis for a forum promoting discussion among community leaders, local government officials, and plant representatives about specific steps that could be taken to encourage source reduction activity and document progress. Such a forum occurred in March, 1988, in West Virginia's Kanawha Valley. The workshop brought together industry leaders, government officials, and citizens to discuss their concerns regarding toxic releases into the local environment. The workshop resulted in an ongoing dialogue among these groups and efforts to create a "scorecard" that local plants could complete to convey their progress to local citizens' groups.

You can also use your findings to influence decisions about pollution permits for the plant you researched. In many cases, you can exercise significant power with state environmental officials when permit decisions are being made. The individual permit writers with whom you established contact during the background information collection phase are your best source of direction here.

Organized campaigns or legal actions directed at facilities found to have poor records are further ways to use findings. You will find the guidebooks mentioned above particularly useful here.

Resources

Citizens Clearinghouse for Hazardous Waste. *Research Guide for Leaders.* (Available from the Citizens Clearinghouse for Hazardous Waste, PO Box 926, Arlington, VA 22216. (703) 276-7070. Price: $3.50.)

How to research opponents, study corporations, analyze communities, and combine research with effective tactics.

Greenpeace and the Environmental Research Foundation (Ben Gordon and Peter Montague). *A Citizen's Toxic Waste Audit Manual.* (Available from Greenpeace USA, 1436 U Street NW, Washington DC 20009. (202) 462-1177. Free.)

Guides citizens on ways to obtain public information, particularly the Toxics Release Inventory database, for specific waste-generating facilities, and then use the data in public campaigns for source reduction of toxic wastes.

INFORM, Inc. (James Cannon). *A Clear View: Guide to Industrial Pollution Control.* New York: INFORM, Inc., 1975. (Available from INFORM, Inc., 381 Park Avenue South, New York, NY 10016. (212) 689-4040. Price: $1.50 plus $2.50 for handling.)

Excellent introduction to environmental issues and laws. Includes a section on gathering and evaluating factory data.

INFORM, Inc. (Catherine G. Miller and Laurence M. Naviasky). *Tracking Toxic Wastes in Ohio.* New York: INFORM, Inc., 1983. (Available from INFORM, Inc., 381 Park Avenue South, New York, NY 10016. (212) 689-4040. Price: $15.00 plus $2.50 for handling; 25 percent discount for nonprofit organizations.)

Provides information on tracking pollution that is also applicable outside Ohio.

National Toxics Campaign. *The Citizens Toxics Protection Manual.* (Available from the National Toxics Campaign, 37 Temple Place, 4th floor, Boston, MA 02111. (617) 482-1477. Price: $35.00.)

Chapters 4 and 5 are excellent resources on researching and approaching a corporation. They are particularly helpful for identifying sources of data and the types of questions to ask about companies.

Chapter 6

Government Policies That Encourage Source Reduction

Your efforts to encourage industrial plants to adopt source reduction as their primary strategy for dealing with toxic wastes can have a major impact on your community. Beyond the immediate local level, however, your successes will be limited by the sheer numbers of industrial facilities and the time constraints facing your group. To achieve a broader impact, you may need to have the support of your state and local officials in promoting source reduction of toxic wastes as government policy.

Government policies and activities can provide significant impetus to corporate progress in three ways: making source reduction a visible priority in the state's environmental protection strategy, getting companies to see more clearly that source reduction is in their own self-interest, and putting corporate practices under a strong public spotlight.

Yet, today, government regulators and industry leaders are generally preoccupied with existing pollution problems and end-of-the-pipe solutions. The congressional Office of Technology Assessment found that well over 99 percent of existing federal environmental resources are spent on cleaning up pollution after it has been created. Waking up your state and federal representatives to the importance of aggressively promoting source reduction will be important.

In 1987, INFORM summarized six specific steps states can take to develop strong source reduction programs (see box: Six Steps States Can Take to Promote Source Reduction). They are not the only legislative approaches possible, but they represent one coherent approach that includes a combination of measures now widely considered in many states. This chapter is designed to give you an understanding of these six steps; INFORM's paper, listed in the references, discusses them in more detail.

Establishing Source Reduction as the Top Priority

The first step for a state is to establish the accepted waste management hierarchy (see Chapter 2) as explicit state policy, distinguishing source reduction from other practices for special emphasis as the top priority. Such a policy counteracts the strong tendency in both industry and government to turn *first* to reactive pollution control solutions to waste problems.

Six Steps States Can Take to Promote Source Reduction

1. Establish the accepted hierarchy of waste management practices as explicit state policy, distinguishing source reduction from other practices for special emphasis as the top priority.

2. Establish a state source reduction office with three distinguishing characteristics.

3. Delegate nine key responsibilities to the office.

4. Enact "right to know" requirements and appoint a "right to know" officer in the state regulatory agency.

5. Take advantage of requirements in other regulatory programs to promote source reduction.

6. Consider instituting source reduction "sticks" to complement the voluntary "carrots" proposed.

As discussed in Chapter 2, the waste management hierarchy specifies source reduction as the first priority, followed in turn by recycling, treatment, and finally landfilling. Federal legislation proposed in Congress contains this policy: "The Congress hereby declares it to be the national policy of the United States that pollution should be prevented or reduced at the source, whenever feasible; pollution that cannot be prevented should be recycled in an environmentally safe manner, whenever feasible; pollution that cannot be prevented or recycled should be treated in an environmentally safe manner; and disposal or other release to the environment should be employed only as a last resort and should be conducted in an environmentally safe manner." Establishing this as state policy as well is an essential step in the creation of a state source reduction program.

Establishing a State Source Reduction Office

The second key step is to establish a state source reduction office with three distinguishing characteristics. These characteristics are necessary if the office is to operate effectively.

1. The office should occupy a high-level niche within the government hierarchy, in, or closely associated with, the leading state environmental agency. This emphasizes the priority the state places on its activities and provides top-level attention to policies designed to stimulate source reduction.

2. The director of the office should be an individual with background in source reduction and industry production processes who commands the respect of corporate leaders — preferably a person with corporate management experience.

3. The office must be separate from, but have some authority over, end-of-the-pipe regulatory offices that deal with a single environmental medium. The office could not fulfill its multimedia, preventive mandate from within a single medium or end-of-the-pipe bureaucracy. Such a setting would narrow its scope and burden its agenda with the daily pressures of meeting regulatory timetables.

Further, direct association with regulatory offices might interfere with the office's ability to create positive relationships with industry.

Delegating Nine Responsibilities to the Office

Once the state source reduction office has been established, it should undertake nine key activities.

1. *Establish goals and timetables for source reduction* by identifying the industries and plants that should receive the highest priority attention from the program and by establishing general numerical goals and realistic timetables for source reduction within these industries. After several years of data analysis, the office should be able to provide the public with clear targets and priorities for source reduction.

2. *Review annual facility-level materials accounts, source reduction plans, and achievements of companies.*

3. *Develop data collection programs* to assure that the state can establish program priorities and evaluate individual and overall progress toward source reduction. At a minimum, data should be collected on the amount and movements of specific polluting chemicals. At the federal level, Title III Section 313 of the Superfund Amendments and Reauthorization Act (SARA) mandates that all plants meeting certain requirements report information on more than 300 toxic chemicals to the annual Toxics Release Inventory (see Appendix A). States could supplement these data with information on other chemicals.

    The key feature of a high-quality database is that it makes it possible to answer the central question: how has the amount of individual toxic pollutants released to the environment changed over a given year and how much of the change was due to source reduction as opposed to declines in production, increased treatment, or other strategies? Measurement of the actual source reduction achieved at a plant thus requires data on the amounts of stored chemicals at the beginning and end of the measurement period, the amounts entering the plant as raw materials, the amounts produced there, the amounts consumed in production, and the amounts leaving as product, as well as the amounts discharged as waste. The state of New Jersey developed an efficient system for collecting this information in 1979.

    It is also vitally important that the data collected be available to the public. *Do not underestimate the importance of a sound public database for driving source reduction.* In an area where standard setting is very difficult, information is one of the most powerful tools for changing industry behavior.

    You have seen that information is your best ammunition when you sit down with local companies. Having hard data on company practices may make all the difference between engaging in a real dialogue with company representatives and just receiving a standard public relations talk.

    Additionally, information publicly identifies — and embarrasses — dirty operators. Companies spend millions of dollars annually on advertising to strengthen their public image and do not want that image threatened by negative publicity demonstrating their lack of concern for the environment. On the other hand, public acknowledgment of those companies that recognize the importance of sound environmental practices and go beyond regulatory requirements with their source reduction efforts might spur competition between companies for future public note.

Most importantly, a good publicly available database will provide you and your state government with a comprehensive understanding of pollution problems and pollution controls in your state. This type of information has been sorely lacking at the state and federal levels; its absence makes creation of sound policy decisions difficult at best.

4. *Identify regulatory and nonregulatory impediments* to source reduction. Often, regulations and loopholes can create inexpensive pollution management alternatives that are dangerous for the environment or public health and that serve to discourage source reduction (**deep well injection**, or disposal of waste deep underground, is one example). State funding priorities, mechanisms, and sources, as well as state tax policies, can also encourage or discourage source reduction. In addition, the office should coordinate activities with the "single medium" offices of the state environmental agency to make sure that source reduction options are considered first.

5. *Analyze findings of source reduction research,* evaluate the need for research on source reduction options within particular industrial processes or categories of chemical pollutants, and establish research programs to meet these needs.

6. *Sponsor meetings* with corporate, trade association, union, environmental, and other leaders to promote source reduction. State officials could use these meetings to promote the company source reduction practices discussed in Chapter 3.

7. *Establish a program of on-site technical assistance* within the government, universities, trade associations, or other organizations to help smaller companies and isolated industries identify source reduction opportunities. This program could provide assistance with source reduction inventories if requested. It could also develop a computerized database in cooperation with other states to share information on specific source reduction technologies and approaches.

8. *Issue biennial reports* on the state's progress in reducing pollution, priority industries for source reduction activities, and the office's activities, achievements, and future goals.

9. *Actively pursue federal funding* for source reduction activities. The Environmental Protection Agency's Pollution Prevention Office currently makes funds available for state technical assistance programs (see Resources). Make sure your state program takes advantage of these funds to build a strong infrastructure that promotes *multimedia* source reduction, and not just compliance strategies for Resource Conservation and Recovery Act regulations.

Establishing "Right to Know" Policies

The state should enact "right to know" requirements and appoint a "right to know" officer in the state regulatory agency. The public needs cooperation from industrial plant managers to obtain the kind of information required to evaluate corporate source reduction strategies. Encourage your state to pass a "sunshine" law, or your governor to issue a "sunshine" proclamation, directing industry to adopt a good neighbor policy and cooperate with the public's efforts to bring source reduction information out into the open. This could be done in several ways.

1. Make it a state policy that industrial plants should provide citizen groups with the data needed to measure source reduction progress and target opportunities. For instance, companies should share information on fluctuations in production levels, amounts of all pollutants leaving the facility, generic types of source reduction used at the company, and the amounts of waste no longer generated. While trade secrets would still be withheld, state policy should condemn irresponsible use of that privilege.

2. Establish statewide good neighbor awards for companies that are cooperative.

3. Fund a "right to know" officer within the state regulatory agency to help citizens obtain data on individual companies, including Toxics Release Inventory information and data on the status of pollution permits given to the company. This officer could also tally information from citizen groups on companies that refuse to work with them at all.

Ensuring All Regulatory Programs Promote Source Reduction

The state's source reduction office should take advantage of requirements in other regulatory programs to promote source reduction. There are a variety of opportunities at both the state and federal level for incorporating a source reduction focus into existing regulatory programs.

1. The office should ensure that personnel in single-medium regulatory programs treat source reduction as the primary pollution management option. For example, it should push for aggressive implementation of the federal 1984 Resource Conservation and Recovery Act (RCRA) amendments that require waste generators to certify that they have a reduction program and to report biennially on their efforts to reduce the volume and/or toxicity of their wastes and the results they have achieved.

2. The office can also provide source reduction training and materials to state and local regulatory personnel responsible for permitting and enforcement. While it may be impractical to develop an independent framework for regulatory source reduction standards, there are many opportunities to weave source reduction options into the fabric of pollution control permits required under existing programs (see, for example, the "Right to Know" section above). Teaching plant inspectors to carry out multimedia inspections, instead of single-medium inspections as they do today, is another key to ensuring that current laws are used to promote source reduction.

3. The office should push for elimination of cheap and dangerous waste disposal activities such as deep well injection of wastes and combustion of toxic wastes as fuel.

4. The office can work to expand the number of chemical substances that are regulated. Approximately 400 wastes or waste streams are regulated under RCRA. The Clean Air Act only regulates six specific

hazardous chemicals and two broad categories at this time (beryllium, mercury, vinyl chloride, asbestos, benzene, arsenic, radionuclides, and coke oven emissions). Under the Clean Water Act, just 126 substances are regulated as "priority pollutants." This is a nearly insignificant number in the face of the 70,000 chemicals used commercially today.

5. Finally, the office should push for stricter enforcement of all environmental laws.

Establishing State Programs Supporting Source Reduction Policies

Beyond encouraging industry to adopt source reduction strategies, the state should consider instituting source reduction "sticks" to complement the voluntary "carrots" proposed. There are several available methods.

1. The state could require any facility requesting permit approval under the Resource Conservation and Recovery Act, Clean Water Act, Clean Air Act, or Toxic Substance Control Act to submit source reduction inventories and plans.

2. The state could require that efforts to meet permitted discharge levels exploit source reduction options before other pollution management techniques.

3. The state could consider imposing a comprehensive pollution fee. In order to promote source reduction, the fee must be applied to the quantity of waste generated prior to any handling or treatment regardless of the manner in which it is handled or the environmental medium into which it is discharged. Pollution must be quantified by the weight of the specific chemicals discharged, not the volume of waste in which the chemicals are contained. Otherwise, activities such as dewatering can reduce the fee applied to a specific pollutant without reducing the amount of the chemicals discharged.

   Such a fee leaves no loopholes because it covers discharges to all media; it is equitable, because industries producing the largest amount of toxic pollutants pay the most money; and it avoids the problems associated with narrower end-of-pipe taxes proposed to date. Using RCRA manifest data as a basis for fees, for example, encourages companies to shift wastes into other routes of discharge rather than reduce the amount of wastes. This also can make state revenue projections unreliable.

   A comprehensive pollution fee structure could involve a number of variations. The fee could vary based upon the level of toxicity of each pollutant, or discharges directly to the environment could involve higher fees than discharges to treatment facilities.

4. The state could consider phasing out or banning particularly offensive chemicals.

   Before you advocate fees or permitting requirements, you might want to consider enlisting your state governor in a cooperative strategy. Organize a meeting of the governor, key industrial representatives, environmentalists, and technical experts in which the governor identifies goals for reductions in pollutants for three or five of the state's most waste-intensive industries. The governor can make a direct, public challenge — with extensive press coverage, if possible — to industry to meet reduction goals in a given time period through its own private strategies, or face fees and/or permit requirements imposed by the state. This strategy allows you to test the spirit of cooperation, and gives industry the opportunity to demonstrate that it can take the initiative. It also may create a political climate conducive to stricter requirements if industry fails to meet the challenge.

Resources

INFORM, Inc. (Dr. Warren R. Muir and Joanna Underwood). *Promoting Hazardous Waste Reduction: Six Steps States Can Take.* New York: INFORM, Inc., 1987. (Available from INFORM, Inc., 381 Park Avenue South, New York, NY 10016. (212) 689-4040. Price: $3.50 plus $2.50 for handling; 25 percent discount for nonprofit organizations.)

This brief guide on state programs was the basis for this chapter.

Local Government Commission, Inc. (Anthony Eulo, Toxics Policy Director). (Suite 203, 909 12th Street, Sacramento, CA 95814. (916) 448-1198.)

Although not discussed in this chapter, promoting source reduction through local government programs has been successful in California where county governments are particularly strong. Anthony Eulo can give you information on local government efforts and how to enhance them.

National Toxics Campaign. *Policy and Program Options for Reduction of Hazardous Waste in Texas.* (Available from the National Toxics Campaign, 37 Temple Place, 4th floor, Boston, MA 02111. (617) 482-1477. Price: $17.00.)

Although this document was prepared specifically for Texas, it contains information of general interest. The section called "Six Aspects of a Comprehensive Waste Reduction Program" is particularly useful and was this guide's sole source for information on integrating inspections and phasing out and banning chemicals.

Office of Technology Assessment. *Serious Reduction of Hazardous Waste.* (Summary available from the Office of Technology Assessment, United States Congress, Washington, DC 20510. (202) 224-8996. Free. Book available from the National Technical Information Service, 5285 Port Royal Road, Springfield, VA 22161. Price: $28.95 in paperback or $6.95 in microfiche, plus $3.00 per order for handling for either edition.)

United States Environmental Protection Agency. Pollution Prevention Office. (EPA Headquarters, (202) 245-4164.)

This office has information on federal funds available for state technical assistance activities.

Ventura County Environmental Health Program. *Hazardous Waste Reduction Guidelines for Environmental Health Programs.* May 1987. (Available from County of Ventura Environmental Health Program, 800 S. Victoria Ave., Ventura, CA 93009. (805) 654-2813. Price: $12.39.)

A useful guide to all aspects of local government programs.

Appendix A

The Toxics Release Inventory

In 1986, the United States Congress passed the Emergency Planning and Community Right to Know Act, or Title III of the Superfund Amendments and Reauthorization Act (SARA). This major law gives you significant new rights to find out about the dangerous chemicals stored, used, and released in your community. In particular, Section 313 of Title III created the Toxics Release Inventory (TRI) to provide public data on "routine" chemical releases from industries across the United States. In the words of New Jersey Governor James Florio, then a congressman, TRI represents the end of "environmental ignorance" for all concerned citizens.

Understanding what TRI is and what information it contains will help you use TRI data most effectively as you work with local companies and state government to maximize source reduction. This appendix gives an overview of TRI and its uses; the article from *Environmental Science and Technology* reprinted at the end of the appendix provides additional details.

TRI in a Nutshell

The TRI legislation requires that all companies meeting certain criteria (see below) report annually to the Environmental Protection Agency how much of each of certain specified chemicals they released to each environmental medium: air, water, and land. The chemicals that must be reported range from acutely lethal to mildly toxic, possibly carcinogenic, or capable of significant adverse effects. (The original TRI chemical list specified 308 individual chemicals and 20 chemical categories. However, this number is constantly changing as chemicals are added to and deleted from the list; currently there are 313 individual chemicals on it.)

The EPA is then required to make this information available to the public. The first inventory, containing 1987 data, was released in mid-1989; it was the first time that specific figures on individual chemical releases from individual industrial plants to individual environmental media were gathered and publicly disseminated.

Which Companies Must Report?

Beginning on July 1, 1988, a company must provide data on its chemical releases to EPA headquarters if it:

- has 10 or more full time employees;

- is a manufacturing industry listed within **Standard Industrial Classification (SIC)** codes 20 through 30 (see Table A-1); and

- manufactures, imports, or processes more than specified amounts of the listed chemicals (50,000 pounds or more in 1988, 25,000 pounds or more each year thereafter), or uses 10,000 pounds or more of one or more of the listed chemicals annually. Tables A-2 and A-3 list the original 308 chemicals and 20 chemical categories (such as cyanide compounds) that companies must report. The chemicals are listed both alphabetically and by **CAS number**, the Chemical Abstract Services number given to each individual chemical by the American Chemical Society in order to uniquely identify chemicals that may have several names.

Table A-1 Industries Required to Report TRI Data

(by Standard Industrial Classification Code)

SIC Codes	Industry Group
2011-2099	Food and kindred products
2111-2141	Tobacco manufacturers
2211-2299	Textile mill products
2311-2399	Apparel and other finished products made from fabrics and other similar materials
2411-2499	Lumber and wood products (except furniture)
2511-2599	Furniture and fixtures
2611-2661	Paper and allied products
2711-2795	Printing, publishing, and allied industries
2811-2899	Chemicals and allied products
2911-2999	Petroleum refining and related industries (e.g., coal)
3011-3079	Rubber and plastic products
3111-3199	Leather and leather products
3211-3299	Stone, clay, glass, and concrete products
3311-3399	Primary metal industries
3411-3499	Fabricated metal products (except machinery and transportation equipment)
3511-3599	Machinery (except electrical)
3611-3699	Electrical and electronic machinery, equipment, and supplies
3711-3799	Transportation equipment
3811-3873	Measuring, analyzing and controlling instruments; photographic, medical and optical goods; watches and clocks
3911-3999	Miscellaneous manufacturing industries

Additional information on SIC codes can be found in the *Standard Industrial Classification Manual 1987*, available from:

National Technical Information Service
5285 Port Royal Road
Springfield, VA 22161
Phone: (703) 487-4650

Table A-2 Chemicals and Chemical Categories that Must Be Reported on TRI (listed alphabetically)

(Original List)

Chemical Name	CAS Number	Chemical Name	CAS Number
Acetaldehyde	75-07-0	Butyl benzyl phthalate	85-68-7
Acetamide	60-35-5	1,2-Butylene oxide	106-88-7
Acetone	67-64-1	Butyraldehyde	123-72-8
Acetonitrile	75-05-8	Cadmium	7440-43-9
2-Acetylaminofluorene	53-96-3	Calcium cyanamide	156-62-7
Acrolein	107-02-8	Captan	133-06-2
Acrylamide	79-06-1	Carbaryl	63-25-2
Acrylic acid	79-10-7	Carbon disulfide	75-15-0
Acrylonitrile	107-13-1	Carbon tetrachloride	56-23-5
Aldrin	309-00-2	Carbonyl sulfide	463-58-1
Allyl chloride	107-05-1	Catechol	120-80-9
Aluminum oxide	1344-28-1	Chloramben	133-90-4
Aluminum (fume or dust)	7429-90-5	Chlordane	57-74-9
2-Aminoanthraquinone	117-79-3	Chlorine	7782-50-5
4-Aminoazobenzene	60-09-3	Chlorine dioxide	10049-04-4
4-Aminobiphenyl	92-67-1	Chloroacetic acid	79-11-8
1-Amino-2-methylanthraquinone	82-28-0	2-Chloroacetophenone	532-27-4
Ammonia	7664-41-7	Chlorobenzene	108-90-7
Ammonium nitrate (solution)	6484-52-2	Chlorobenzilate	510-15-6
Ammonium sulfate (solution)	7783-20-2	Chloroform	67-66-3
Aniline	62-53-3	Chloromethyl methyl ether	107-30-2
o-Anisidine hydrochloride	134-29-2	Chloroprene	126-99-8
o-Anisidine	90-04-0	Chlorothalonil	1897-45-6
p-Anisidine	104-94-9	Chromium	7440-47-3
Anthracene	120-12-7	Cobalt	7440-48-4
Antimony	7440-36-0	Copper	7440-50-8
Arsenic	7440-38-2	p-Cresidine	120-71-8
Asbestos (friable)	1332-21-4	m-Cresol	108-39-4
Auramine	492-80-8	o-Cresol	95-48-7
Barium	7440-39-3	p-Cresol	106-44-5
Benzal chloride	98-87-3	Cresol (mixed isomers)	1319-77-3
Benzamide	55-21-0	Cumene hydroperoxide	80-15-9
Benzene	71-43-2	Cupferron	135-20-6
Benzidine	92-87-5	Cyclohexane	110-82-7
p-Benzoquinone	106-51-4	C.I. Acid Blue 9, diammonium salt	2650-18-2
Benzotrichloride	98-07-7	C.I. Acid Blue 9, disodium salt	3844-45-9
Benzoyl chloride	98-88-4	C.I. Acid Green 3	4680-78-8
Benzoyl peroxide	94-36-0	C.I. Basic Green 4	569-64-2
Benzyl chloride	100-44-7	C.I. Basic Red 1	989-38-8
Beryllium	7440-41-7	C.I. Direct Black 38	1937-37-7
Biphenyl	92-52-4	C.I. Direct Blue 6	2602-46-2
Bis(2-ethylhexyl) adipate	103-23-1	C.I. Direct Brown 95	16071-86-6
Bromoform	75-25-2	C.I. Disperse Yellow 3	2832-40-8
1,3-Butadiene	106-99-0	C.I. Food Red 15	81-88-9
1-Butanol	71-36-3	C.I. Food Red 5	3761-53-3
Butyl acrylate	141-32-2	C.I. Solvent Orange 7	3118-97-6
tert-Butyl alcohol	75-65-0	C.I. Solvent Yellow 14	842-07-9

Table A-2 (continued) TRI Chemicals (listed alphabetically)

Chemical Name	CAS Number	Chemical Name	CAS Number
C.I. Solvent Yellow 3	97-56-3	Epichlorohydrin	106-89-8
C.I. Vat Yellow 4	128-66-5	Ethyl acrylate	140-88-5
2,4-D	94-75-7	Ethyl carbamate (urethane)	51-79-6
Decabromodiphenyl oxide	1163-19-5	Ethyl chloride	75-00-3
Diallate	2303-16-4	Ethyl chloroformate	541-41-3
2,4-Diaminoanisole sulfate	39156-41-7	Ethylbenzene	100-41-4
2,4-Diaminoanisole	615-05-4	Ethylene	74-85-1
4,4'-Diaminodiphenyl ether	101-80-4	Ethylene dibromide	106-93-4
2,4-Diaminotoluene	95-80-7	Ethylene dichloride	107-06-2
Diazomethane	334-88-3	Ethylene glycol	107-21-1
Dibenzofuran	132-64-9	Ethylene glycol monoethyl ether	110-80-5
1,2-Dibromo-3-chloropropane	96-12-8	Ethylene oxide	75-21-8
Dibutyl phthalate	84-74-2	Ethyleneimine	151-56-4
m-Dichlorobenzene	541-73-1	Ethylenethiourea	96-45-7
o-Dichlorobenzene	95-50-1	Fluometuron	2164-17-2
p-Dichlorobenzene	106-46-7	Formaldehyde	50-00-0
Dichlorobenzene	25321-22-6	Freon 113	76-13-1
3,3'-Dichlorobenzidine	91-94-1	Heptachlor	76-44-8
Dichlorobromomethane	75-27-4	Hexachlorobenzene	118-74-1
Dichloroethyl ether	111-44-4	Hexachlorobutadiene	87-68-3
1,1-Dichloroethylene	75-35-4	Hexachlorocyclopentadiene	77-47-4
1,2-Dichloroethylene	540-59-0	Hexachloroethane	67-72-1
Dichloroisopropyl ether	108-60-1	Hexachloronaphthalene	1335-87-1
Dichloromethyl ether	542-88-1	Hexamethylphosphoramide	680-31-9
2,4-Dichlorophenol	120-83-2	Hydrazine	302-01-2
1,3-Dichloropropene	542-75-6	Hydrazine sulfate	10034-93-2
Dichlorvos	62-73-7	Hydrochloric acid	7647-01-0
Dicofol	115-32-2	Hydrogen cyanide	74-90-8
1:2,3:4-Diepoxybutane	1464-53-5	Hydrogen fluoride	7664-39-3
Diethanolamine	111-42-2	Hydroquinone	123-31-9
Diethyl sulfate	64-67-5	Isobutyraldehyde	78-84-2
Diethylhexyl phthalate	117-81-7	Isopropyl alcohol	67-63-0
Diethylphthalate	84-66-2	4,4'-Isopropylidenediphenol	80-05-7
3,3'-Dimethoxybenzidine	119-90-4	Lead	7439-92-1
Dimethyl sulfate	77-78-1	Lindane	58-89-9
4-Dimethylaminoazobenzene	60-11-7	Maleic anhydride	108-31-6
N,N-Dimethylaniline	121-69-7	Maneb	12427-38-2
3,3'-Dimethylbenzidine	119-93-7	Manganese	7439-96-5
Dimethylcarbomyl chloride	79-44-7	MBI	101-68-8
1,1-Dimethylhydrazine	57-14-7	Melamine	108-78-1
2,4-Dimethylphenol	105-67-9	Mercury	7439-97-6
Dimethylphthalate	131-11-3	Methanol	67-56-1
2,4-Dinitrophenol	51-28-5	Methoxychlor	72-43-5
2,4-Dinitrotoluene	121-14-2	2-Methoxyethanol	109-86-4
2,6-Dinitrotoluene	606-20-2	Methyl acrylate	96-33-3
4,6-Dinitro-o-cresol	534-52-1	Methyl bromide	74-83-9
Dioxane	123-91-1	Methyl chloride	74-87-3
1,2-Diphenylhydrazine	122-66-7	1-Methyl ethyl benzene (Cumene)	98-82-8
Di-n-octylphthalate	117-84-0	Methyl ethyl ketone (MEK)	78-93-3
Di-n-propylnitrosamine	621-64-7	Methyl hydrazine	60-34-4

Table A-2 (continued) TRI Chemicals (listed alphabetically)

Chemical Name	CAS Number	Chemical Name	CAS Number
Methyl iodide	74-88-4	Phosphoric acid	7664-38-2
Methyl isobutyl ketone	108-10-1	Phosphorous (yellow or white)	7723-14-0
Methyl isocyanate	624-83-9	Phthalic anhydride	85-44-9
Methyl methacrylate	80-62-6	Picric acid	88-89-1
Methyl tert-butyl ether	1634-04-4	1,3-Propane sultone	1120-71-4
Methylene bromide	74-95-3	beta-Propiolactone	57-57-8
Methylene chloride	75-09-2	Propionaldehyde	123-38-6
4,4'-Methylenebis(2-chloroaniline)	101-14-4	Propxur	114-26-1
4,4'-Methylenebis(N,N-dimethyl)	101-61-1	Propylene	115-07-1
4,4'-Methylenedianiline	101-77-9	Propylene dichloride	78-87-5
Michler's ketone	90-94-8	Propylene oxide	75-56-9
Molybdenum trioxide	1313-27-5	Propyleneimine	75-55-8
Mustard gas	505-60-2	Pyridine	110-86-1
Naphthalene	91-20-3	Quinoline	91-22-5
alpha-Naphthylamine	134-32-7	Saccharin	81-07-2
beta-Naphthylamine	91-59-8	Safrole	94-59-7
Nickel	7440-02-0	sec-Butyl alcohol	78-92-2
Nitric acid	7697-37-2	Selenium	7782-49-2
Nitrilotriacetic acid	139-13-9	Silver	744022-4
Nitrobenzene	98-95-3	Sodium hydroxide (solution)	1310-73-2
4-Nitrobiphenyl	92-93-3	Sodium sulfate (solution)	7757-82-6
Nitrofen	1836-75-5	Styrene	100-42-5
Nitrogen mustard	51-75-2	Styrene oxide	96-09-3
Nitroglycerin	55-63-0	Sulfuric acid	7664-93-9
p-Nitrophenol	100-02-7	Terephthalic acid	100-21-0
2-Nitrophenol	88-75-5	1,1,2,2-Tetrachloroethane	79-34-5
2-Nitropropane	79-46-9	Tetrachloroethylene	127-18-4
N-Nitrosodiethylamine	55-18-5	Tetrachlorvinphos	961-11-5
N-Nitrosodimethylamine	62-75-9	Thallium	7440-28-0
N-Nitrosodiphenylamine	86-30-6	Thioacetamide	62-55-5
p-Nitrosodiphenylamine	156-10-5	4,4'-Thiodianiline	139-65-1
N-Nitrosodi-n-butylamine	924-16-3	Thiourea	62-56-6
N-Nitrosomethylvinylamine	4549-40-0	Thorium dioxide	1314-20-1
N-Nitrosomorpholine	59-89-2	Titanium dioxide	13463-67-7
N-Nitrosonornicotine	16543-55-8	Titanium tetrachloride	7550-45-0
N-Nitrosopiperidine	100-75-4	Toluene	108-88-3
N-Nitroso-N-ethyl urea	759-73-9	Toluenediamine	25376-45-8
N-Nitroso-N-methylurea	684-93-5	Toluene-2,4-diisocyanate	584-84-9
5-Nitro-o-anisidine	99-59-2	Toluene-2,6-diisocyanate	91-08-7
Octachloronaphthalene	2234-13-1	o-Toluidine hydrochloride	636-21-5
Osmium tetroxide	20816-12-0	o-Toluidine	95-53-4
Parathion	56-38-2	Toxaphene	8001-35-2
PCBs	1336-36-3	Triaziquone	68-76-8
Pentachloronitrobenzene (PCNB)	82-68-8	Trichlorfon	52-68-6
Pentachlorophenol	87-86-5	1,2,4-Trichlorobenzene	120-82-1
Peracetic acid	79-21-0	1,1,1-Trichloroethane	71-55-6
Phenol	108-95-2	1,1,2-Trichloroethane	79-00-5
p-Phenylenediamine	106-50-3	Trichloroethylene	79-01-6
2-Phenylphenol	90-43-7	2,4,5-Trichlorophenol	95-95-4
Phosgene	75-44-5	2,4,6-Trichlorophenol	88-06-2

Table A-2 (continued) TRI Chemicals (listed alphabetically)

Chemical Name	CAS Number	Chemical Name	CAS Number
Trifluralin	1582-09-8	Beryllium compounds, NOS*	. . . .
1,2,4-Trimethylbenzene	95-63-6	Cadmium compounds, NOS*	. . . .
Tris(2,3-dibromopropyl) phosphate	126-72-7	Chlorinated phenol, NOS*	. . . .
Vanadium (fume or dust)	7440-62-2	Chromium compounds, NOS*	. . . .
Vinyl acetate	108-05-4	Cobalt compounds	. . . .
Vinyl bromide	593-60-2	Copper compounds	. . . .
Vinyl chloride	75-01-4	Cyanide compounds	. . . .
m-Xylene	108-38-3	Glycol ethers	. . . .
o-Xylene	95-47-6	Lead compounds, NOS*	. . . .
p-Xylene	106-42-3	Manganese compounds	. . . .
Xylene (mixed isomers)	1330-20-7	Mercury compounds, NOS*	. . . .
2,6-Xylidine	87-62-7	Nickel compounds, NOS*	. . . .
Zinc (fume or dust)	7440-66-6	Polybrominated biphenyls, NOS*	. . . .
Zineb	12122-67-7	Selenium compounds, NOS*	. . . .
Antimony compounds, NOS*	. . . .	Silver compounds, NOS*	. . . .
Arsenic compounds, NOS*	. . . .	Thallium compounds, NOS*	. . . .
Barium compounds, NOS*	. . . .	Zinc compounds, NOS*	. . . .

(NOS* equals "Not Otherwise Specified")

(Changes from Original List)

(Chemicals Added on December 1, 1989)

Chemical Name	CAS Number
Allyl alcohol	107-18-6
Creosote	8001-58-9
2,3-Dichloropropane	78-88-6
m-Dinitrobenzenol	99-65-0
o-Dinitrobenzenol	528-29-0
p-Dinitrobenzenol	100-25-4
Dinitrotoluene (mixed isomers)	25321-14-6
Isosafrole	120-58-1
Toluene diisocyanate	26471-62-5

(Chemicals Deleted from Original List)

Chemical Name	CAS Number
C.I. Acid Blue 9 diammonium salt	2650-18-2
C.I. Acid Blue 9 disodium salt	3844-45-9
Melamine	108-77-1
Sodium hydroxide solution	1310-73-2
Sodium sulfate	7757-82-6
Titanium dioxide	13463-67-7

Table A-3 Chemicals and Chemical Categories that Must Be Reported on TRI (listed by CAS number)

(Original List)

Chemical Name	CAS Number	Chemical Name	CAS Number
Formaldehyde	50-00-0	Methylene chloride	75-09-2
2,4-Dinitrophenol	51-28-5	Carbon disulfide	75-15-0
Nitrogen mustard	51-75-2	Ethylene oxide	75-21-8
Ethyl carbamate (urethane)	51-79-6	Bromoform	75-25-2
Trichlorfon	52-68-6	Dichlorobromomethane	75-27-4
2-Acetylaminofluorene	53-96-3	1,1-Dichloroethylene	75-35-4
N-Nitrosodiethylamine	55-18-5	Phosgene	75-44-5
Benzamide	55-21-0	Propyleneimine	75-55-8
Nitroglycerin	55-63-0	Propylene oxide	75-56-9
Carbon tetrachloride	56-23-5	tert-Butyl alcohol	75-65-0
Parathion	56-38-2	Freon 113	76-13-1
1,1-Dimethylhydrazine	57-14-7	Heptachlor	76-44-8
beta-Propiolactone	57-57-8	Hexachlorocyclopentadiene	77-47-4
Chlordane	57-74-9	Dimethyl sulfate	77-78-1
Lindane	58-89-9	Isobutyraldehyde	78-84-2
N-Nitrosomorpholine	59-89-2	Propylene dichloride	78-87-5
4-Aminoazobenzene	60-09-3	sec-Butyl alcohol	78-92-2
4-Dimethylaminoazobenzene	60-11-7	Methyl ethyl ketone (MEK)	78-93-3
Methyl hydrazine	60-34-4	1,1,2-Trichloroethane	79-00-5
Acetamide	60-35-5	Trichloroethylene	79-01-6
Aniline	62-53-3	Acrylamide	79-06-1
Thioacetamide	62-55-5	Acrylic acid	79-10-7
Thiourea	62-56-6	Chloroacetic acid	79-11-8
Dichlorvos	62-73-7	Peracetic acid	79-21-0
N-Nitrosodimethylamine	62-75-9	1,1,2,2-Tetrachloroethane	79-34-5
Carbaryl	63-25-2	Dimethylcarbomyl chloride	79-44-7
Diethyl sulfate	64-67-5	2-Nitropropane	79-46-9
Methanol	67-56-1	4,4'-Isopropylidenediphenol	80-05-7
Isopropyl alcohol	67-63-0	Cumene hydroperoxide	80-15-9
Acetone	67-64-1	Methyl methacrylate	80-62-6
Chloroform	67-66-3	Saccharin	81-07-2
Hexachloroethane	67-72-1	C.I. Food Red 15	81-88-9
Triaziquone	68-76-8	1-Amino-2-methylanthraquinone	82-28-0
1-Butanol	71-36-3	Pentachloronitrobenzene (PCNB)	82-68-8
Benzene	71-43-2	Diethylphthalate	84-66-2
1,1,1-Trichloroethane	71-55-6	Dibutyl phthalate	84-74-2
Methoxychlor	72-43-5	Phthalic anhydride	85-44-9
Methyl bromide	74-83-9	Butyl benzyl phthalate	85-68-7
Ethylene	74-85-1	N-Nitrosodiphenylamine	86-30-6
Methyl chloride	74-87-3	2,6-Xylidine	87-62-7
Methyl iodide	74-88-4	Hexachlorobutadiene	87-68-3
Hydrogen cyanide	74-90-8	Pentachlorophenol	87-86-5
Methylene bromide	74-95-3	2,4,6-Trichlorophenol	88-06-2
Ethyl chloride	75-00-3	2-Nitrophenol	88-75-5
Vinyl chloride	75-01-4	Picric acid	88-89-1
Acetonitrile	75-05-8	o-Anisidine	90-04-0
Acetaldehyde	75-07-0	2-Phenylphenol	90-43-7

Table A-3 (continued) TRI Chemicals (by CAS number)

Chemical Name	CAS Number	Chemical Name	CAS Number
Michler's ketone	90-94-8	1,2-Butylene oxide	106-88-7
Toluene-2,6-diisocyanate	91-08-7	Epichlorohydrin	106-89-8
Naphthalene	91-20-3	Ethylene dibromide	106-93-4
Quinoline	91-22-5	1,3-Butadiene	106-99-0
beta-Naphthylamine	91-59-8	Acrolein	107-02-8
3,3'-Dichlorobenzidine	91-94-1	Allyl chloride	107-05-1
Biphenyl	92-52-4	Ethylene dichloride	107-06-2
4-Aminobiphenyl	92-67-1	Acrylonitrile	107-13-1
Benzidine	92-87-5	Ethylene glycol	107-21-1
4-Nitrobiphenyl	92-93-3	Chloromethyl methyl ether	107-30-2
Benzoyl peroxide	94-36-0	Vinyl acetate	108-05-4
Safrole	94-59-7	Methyl isobutyl ketone	108-10-1
2,4-D	94-75-7	Maleic anhydride	108-31-6
o-Xylene	95-47-6	m-Xylene	108-38-3
o-Cresol	95-48-7	m-Cresol	108-39-4
o-Dichlorobenzene	95-50-1	Dichloroisopropyl ether	108-60-1
o-Toluidine	95-53-4	Melamine	108-78-1
1,2,4-Trimethylbenzene	95-63-6	Toluene	108-88-3
2,4-Diaminotoluene	95-80-7	Chlorobenzene	108-90-7
2,4,5-Trichlorophenol	95-95-4	Phenol	108-95-2
Styrene oxide	96-09-3	2-Methoxyethanol	109-86-4
1,2-Dibromo-3-chloropropane	96-12-8	Ethylene glycol monoethyl ether	110-80-5
Methyl acrylate	96-33-3	Cyclohexane	110-82-7
Ethylenethiourea	96-45-7	Pyridine	110-86-1
C.I. Solvent Yellow 3	97-56-3	Diethanolamine	111-42-2
Benzotrichloride	98-07-7	Dichloroethyl ether	111-44-4
1-Methyl ethyl benzene (Cumene)	98-82-8	Propxur	114-26-1
Benzal chloride	98-87-3	Propylene	115-07-1
Benzoyl chloride	98-88-4	Dicofol	115-32-2
Nitrobenzene	98-95-3	2-Aminoanthraquinone	117-79-3
5-Nitro-o-anisidine	99-59-2	Diethylhexyl phthalate	117-81-7
p-Nitrophenol	100-02-7	Di-n-octylphthalate	117-84-0
Terephthalic acid	100-21-0	Hexachlorobenzene	118-74-1
Ethylbenzene	100-41-4	3,3'-Dimethoxybenzidine	119-90-4
Styrene	100-42-5	3,3'-Dimethylbenzidine	119-93-7
Benzyl chloride	100-44-7	Anthracene	120-12-7
N-Nitrosopiperidine	100-75-4	p-Cresidine	120-71-8
4,4'-Methylenebis (2-chloroaniline)	101-14-4	Catechol	120-80-9
4,4'-Methylenebis (N,N-dimethyl)	101-61-1	1,2,4-Trichlorobenzene	120-82-1
MBI	101-68-8	2,4-Dichlorophenol	120-83-2
4,4'-Methylenedianiline	101-77-9	2,4-Dinitrotoluene	121-14-2
4,4'-Diaminodiphenyl ether	101-80-4	N,N-Dimethylaniline	121-69-7
Bis (2-ethylhexyl) adipate	103-23-1	1,2-Diphenylhydrazine	122-66-7
p-Anisidine	104-94-9	Hydroquinone	123-31-9
2,4-Dimethylphenol	105-67-9	Propionaldehyde	123-38-6
p-Xylene	106-42-3	Butyraldehyde	123-72-8
p-Cresol	106-44-5	Dioxane	123-91-1
p-Dichlorobenzene	106-46-7	Tris (2,3-dibromopropyl) phosphate	126-72-7
p-Phenylenediamine	106-50-3	Chloroprene	126-99-8
p-Benzoquinone	106-51-4	Tetrachloroethylene	127-18-4

Table A-3 (continued) TRI Chemicals (by CAS number)

Chemical Name	CAS Number	Chemical Name	CAS Number
C.I. Vat Yellow 4	128-66-5	Xylene (mixed isomers)	1330-20-7
Dimethylphthalate	131-11-3	Asbestos (friable)	1332-21-4
Dibenzofuran	132-64-9	Hexachloronaphthalene	1335-87-1
Captan	133-06-2	PCBs	1336-36-3
Chloramben	133-90-4	Aluminum oxide	1344-28-1
o-Anisidine hydrochloride	134-29-2	1:2,3:4-Diepoxybutane	1464-53-5
alpha-Naphthylamine	134-32-7	Trifluralin	1582-09-8
Cupferron	135-20-6	Methyl tert-butyl ether	1634-04-4
Nitrilotriacetic acid	139-13-9	Nitrofen	1836-75-5
4,4'-Thiodianiline	139-65-1	Chlorothalonil	1897-45-6
Ethyl acrylate	140-88-5	C.I. Direct Black 38	1937-37-7
Butyl acrylate	141-32-2	Fluometuron	2164-17-2
Ethyleneimine	151-56-4	Octachloronaphthalene	2234-13-1
p-Nitrosodiphenylamine	156-10-5	Diallate	2303-16-4
Calcium cyanamide	156-62-7	C.I. Direct Blue 6	2602-46-2
Hydrazine	302-01-2	C.I. Acid Blue 9, diammonium salt	2650-18-2
Aldrin	309-00-2	C.I. Disperse Yellow 3	2832-40-8
Diazomethane	334-88-3	C.I. Solvent Orange 7	3118-97-6
Carbonyl sulfide	463-58-1	C.I. Food Red 5	3761-53-3
Auramine	492-80-8	C.I. Acid Blue 9, disodium salt	3844-45-9
Mustard gas	505-60-2	N-Nitrosomethylvinylamine	4549-40-0
Chlorobenzilate	510-15-6	C.I. Acid Green 3	4680-78-8
2-Chloroacetophenone	532-27-4	Ammonium nitrate (solution)	6484-52-2
4,6-Dinitro-o-cresol	534-52-1	Aluminum (fume or dust)	7429-90-5
1,2-Dichloroethylene	540-59-0	Lead	7439-92-1
Ethyl chloroformate	541-41-3	Manganese	7439-96-5
m-Dichlorobenzene	541-73-1	Mercury	7439-97-6
1,3-Dichloropropene	542-75-6	Nickel	7440-02-0
Dichloromethyl ether	542-88-1	Silver	7440-22-4
C.I. Basic Green 4	569-64-2	Thallium	7440-28-0
Toluene-2,4-diisocyanate	584-84-9	Antimony	7440-36-0
Vinyl bromide	593-60-2	Arsenic	7440-38-2
2,6-Dinitrotoluene	606-20-2	Barium	7440-39-3
2,4-Diaminoanisole	615-05-4	Beryllium	7440-41-7
Di-n-propylnitrosamine	621-64-7	Cadmium	7440-43-9
Methyl isocyanate	624-83-9	Chromium	7440-47-3
o-Toluidine hydrochloride	636-21-5	Cobalt	7440-48-4
Hexamethylphosphoramide	680-31-9	Copper	7440-50-8
N-Nitroso-N-methylurea	684-93-5	Vanadium (fume or dust)	7440-62-2
N-Nitroso-N-ethyl urea	759-73-9	Zinc (fume or dust)	7440-66-6
C.I. Solvent Yellow 14	842-07-9	Titanium tetrachloride	7550-45-0
N-Nitrosodi-n-butylamine	924-16-3	Hydrochloric acid	7647-01-0
Tetrachlorvinphos	961-11-5	Phosphoric acid	7664-38-2
C.I. Basic Red 1	989-38-8	Hydrogen fluoride	7664-39-3
1,3-Propane sultone	1120-71-4	Ammonia	7664-41-7
Decabromodiphenyl oxide	1163-19-5	Sulfuric acid	7664-93-9
Sodium hydroxide (solution)	1310-73-2	Nitric acid	7697-37-2
Molybdenum trioxide	1313-27-5	Phosphorous (yellow or white)	7723-14-0
Thorium dioxide	1314-20-1	Sodium sulfate (solution)	7757-82-6
Cresol (mixed isomers)	1319-77-3	Selenium	7782-49-2

Table A-3 (continued) TRI Chemicals (by CAS number)

Chemical Name	CAS Number	Chemical Name	CAS Number
Chlorine	7782-50-5	Beryllium compounds, NOS*	. . . .
Ammonium sulfate (solution)	7783-20-2	Cadmium compounds, NOS*	. . . .
Toxaphene	8001-35-2	Chlorinated phenol, NOS*	. . . .
Hydrazine sulfate	10034-93-2	Chromium compounds, NOS*	. . . .
Chlorine dioxide	10049-04-4	Cobalt compounds	. . . .
Zineb	12122-67-7	Copper compounds	. . . .
Maneb	12427-38-2	Cyanide compounds	. . . .
Titanium dioxide	13463-67-7	Glycol ethers	. . . .
C.I. Direct Brown 95	16071-86-6	Lead compounds, NOS*	. . . .
N-Nitrosonornicotine	16543-55-8	Manganese compounds	. . . .
Osmium tetroxide	20816-12-0	Mercury compounds, NOS*	. . . .
Dichlorobenzene, NOS*	25321-22-6	Nickel compounds, NOS*	. . . .
Toluenediamine	25376-45-8	Polybrominated biphenyls, NOS*	. . . .
2,4-Diaminoanisole sulfate	39156-41-7	Selenium compounds, NOS*	. . . .
Antimony compounds, NOS*	. . . .	Silver compounds, NOS*	. . . .
Arsenic compounds, NOS*	. . . .	Thallium compounds, NOS*	. . . .
Barium compounds, NOS*	. . . .	Zinc compounds, NOS*	. . . .

(NOS* equals "Not Otherwise Specified")

(Changes from Original List)

(Chemicals Added on December 1, 1989)

Chemical Name	CAS Number
2,3-Dichloropropane	78-88-6
m-Dinitrobenzenol	99-65-0
p-Dinitrobenzenol	100-25-4
Allyl alcohol	107-18-6
Isosafrole	120-58-1
o-Dinitrobenzenol	528-29-0
Creosote	8001-58-9
Dinitrotoluene (mixed isomers)	25321-14-6
Toluene diisocyanate	26471-62-5

(Chemicals Deleted from Original List)

Chemical Name	CAS Number
Melamine	108-77-1
Sodium hydroxide solution	1310-73-2
C.I. Acid Blue 9 diammonium salt	2650-18-2
C.I. Acid Blue 9 disodium salt	3844-45-9
Sodium sulfate	7757-82-6
Titanium dioxide	13463-67-7

What Information Is Reported?

Companies reporting Toxics Release Inventory data fill out EPA **Form R**, the Toxic Chemical Release Reporting Form, for each chemical (see blank form, pages 68-72). It asks for information on the company and its chemical releases.

- Name, location, and type of business.

- Name and telephone number of public contact person at the company.

- National Pollution Discharge Elimination System (NPDES) and Resource Conservation and Recovery Act (RCRA) permit numbers. (These pertain to federal and state permits for surface water emissions and land disposal, respectively; you can use these numbers to request information from state and federal regulatory agencies about specific permitted activities at the facility.)

- Surface waters receiving TRI chemical pollutants.

- Name and location of publicly owned treatment works (POTW) and off-site disposal facilities to which waste from the company is transported.

- Use of the chemical.

- Maximum amounts of the chemical on-site at any time during the preceding year.

- Quantity of the chemical released as fugitive air emissions, stack emissions, discharges to water, underground injection, land disposal, discharges to a POTW, and discharges to other off-site location.

- Type and efficiency of treatment methods employed.

- Technique for estimating releases and treatment efficiency. (The company selects among codes for four estimate techniques: actual monitoring of releases, mass balance calculations, generic mathematical derivations for emissions factors, and other.)

- Optional information on the type and amount of waste minimization achieved at the facility, a production index to indicate whether production has increased or decreased at the facility, and an indication of why the waste minimization action was taken.

The EPA has produced *A Guide to the Toxic Chemical Release Inventory Form R* (see Resources); it contains a section-by-section description of data collected on the form. This booklet will be extremely useful to you if you try to read a TRI form.

How To Get TRI Information

There are several ways you can obtain TRI information, with or without computers. The EPA, required by law to make the data publicly accessible by telecommunications, has contracted to do so through the National Library of Medicine's TOXNET system. The system is available 24 hours a day, seven days a week, to anyone with a computer and a modem.

(Important: Type or print; read instructions before completing form.)

♻EPA U.S. Environmental Protection Agency

TOXIC CHEMICAL RELEASE INVENTORY REPORTING FORM
Section 313 of the Emergency Planning and Community Right-to-Know Act of 1986, also known as Title III of the Superfund Amendments and Reauthorization Act

EPA FORM R	PART I. FACILITY IDENTIFICATION INFORMATION	(This space for your optional use.)

Public reporting burden for this collection of information is estimated to vary from 30 to 34 hours per response, with an average of 32 hours per response, including time for reviewing instructions, searching existing data sources, gathering and maintaining the data needed, and completing and reviewing the collection of information. Send comments regarding this burden estimate or any other aspect of this collection of information, including suggestions for reducing this burden, to Chief, Information Policy Branch (PM-223), US EPA, 401 M St., SW, Washington, D.C. 20460 Attn: TRI Burden and to the Office of Information and Regulatory Affairs, Office of Management and Budget Paperwork Reduction Project (2070-0093), Washington, D.C. 20603.

1.

1.1 Are you claiming the chemical identity on page 3 trade secret?

[ ] Yes (Answer question 1.2; Attach substantiation forms.) [ ] No (Do not answer 1.2; Go to question 1.3.)

1.2 If "Yes" in 1.1, is this copy:

[ ] Sanitized [ ] Unsanitized

1.3 Reporting Year

19____

2. CERTIFICATION (Read and sign after completing all sections.)

I hereby certify that I have reviewed the attached documents and that, to the best of my knowledge and belief, the submitted information is true and complete and that the amounts and values in this report are accurate based on reasonable estimates using data available to the preparers of this report.

Name and official title of owner/operator or senior management official

Signature		Date signed

3. FACILITY IDENTIFICATION

3.1

Facility or Establishment Name

Street Address

City	County
State	Zip Code

TRI Facility Identification Number

WHERE TO SEND COMPLETED FORMS:

1. EPCRA REPORTING CENTER
P.O. BOX 23779
WASHINGTON, DC 20026-3779
ATTN: TOXIC CHEMICAL RELEASE INVENTORY

2. APPROPRIATE STATE OFFICE (See instructions in Appendix G)

3.2 This report contains information for (Check only one):   a. [ ] An entire facility   b. [ ] Part of a facility.

3.3 Technical Contact | Telephone Number (include area code)

3.4 Public Contact | Telephone Number (include area code)

3.5 SIC Code (4 digit)

a.	b.	c.	d.	e.	f.

3.6

Latitude			Longitude		
Degrees.	Minutes	Seconds	Degrees	Minutes	Seconds

3.7 Dun & Bradstreet Number(s)

a.	b.

3.8 EPA Identification Number(s) (RCRA I.D. No.)

a.	b.

3.9 NPDES Permit Number(s)

a.	b.

3.10 Receiving Streams or Water Bodies (enter one name per box)

a.	b.
c.	d.
e.	f.

3.11 Underground Injection Well Code (UIC) Identification Number(s)

a.	b.

4. PARENT COMPANY INFORMATION

4.1 Name of Parent Company | **4.2** Parent Company's Dun & Bradstreet Number

EPA Form 9350-1 (1-90) Revised – Do not use previous versions.

<table>
<tr><td rowspan="2">&EPA</td><td>EPA FORM R
PART II. OFF-SITE LOCATIONS TO WHICH TOXIC
CHEMICALS ARE TRANSFERRED IN WASTES</td><td>(This space for your optional use.)</td></tr>
</table>

1. PUBLICLY OWNED TREATMENT WORKS (POTWs)

1.1 POTW name		1.2 POTW name	
Street Address		Street Address	
City	County	City	County
State	Zip	State	Zip

2. OTHER OFF-SITE LOCATIONS (DO NOT REPORT LOCATIONS TO WHICH WASTES ARE SENT ONLY FOR RECYCLING OR REUSE).

2.1 Off-site location name		2.2 Off-site location name	
EPA Identification Number (RCRA ID. No.)		EPA Identification Number (RCRA ID. No.)	
Street Address		Street Address	
City	County	City	County
State	Zip	State	Zip
Is location under control of reporting facility or parent company? [ ]Yes [ ]No		Is location under control of reporting facility or parent company? [ ]Yes [ ]No	

2.3 Off-site location name		2.4 Off-site location name	
EPA Identification Number (RCRA ID. No.)		EPA Identification Number (RCRA ID. No.)	
Street Address		Street Address	
City	County	City	County
State	Zip	State	Zip
Is location under control of reporting facility or parent company? [ ]Yes [ ]No		Is location under control of reporting facility or parent company? [ ]Yes [ ]No	

2.5 Off-site location name		2.6 Off-site location name	
EPA Identification Number (RCRA ID. No.)		EPA Identification Number (RCRA ID. No.)	
Street Address		Street Address	
City	County	City	County
State	Zip	State	Zip
Is location under control of reporting facility or parent company? [ ]Yes [ ]No		Is location under control of reporting facility or parent company? [ ]Yes [ ]No	

[ ] Check if additional pages of Part II are attached. How many? _____

EPA Form 9350-1 (1-90) Revised - Do not use previous versions.

⊕EPA	EPA FORM **R** PART III. CHEMICAL–SPECIFIC INFORMATION	(This space for your optional use.)

1. CHEMICAL IDENTITY (Do not complete this section if you complete Section 2.)

1.1	[Reserved]
1.2	CAS Number (Enter only one number exactly as it appears on the 313 list. Enter NA if reporting a chemical category.)
1.3	Chemical or Chemical Category Name (Enter only one name exactly as it appears on the 313 list.)
1.4	Generic Chemical Name (Complete only if Part I, Section 1.1 is checked "Yes." Generic name must be structurally descriptive.)

2.	MIXTURE COMPONENT IDENTITY (Do not complete this section if you complete Section 1.) Generic Chemical Name Provided by Supplier (Limit the name to a maximum of 70 characters (e.g., numbers, letters, spaces, punctuation).)

3. ACTIVITIES AND USES OF THE CHEMICAL AT THE FACILITY (Check all that apply.)

3.1	Manufacture the chemical:	a. [ ] Produce b. [ ] Import	If produce or import: c. [ ] For on-site use/processing e. [ ] As a byproduct	d. [ ] For sale/distribution f. [ ] As an impurity
3.2	Process the chemical:	a. [ ] As a reactant d. [ ] Repackaging only	b. [ ] As a formulation component	c. [ ] As an article component
3.3	Otherwise use the chemical:	a. [ ] As a chemical processing aid	b. [ ] As a manufacturing aid	c. [ ] Ancillary or other use

4. MAXIMUM AMOUNT OF THE CHEMICAL ON-SITE AT ANY TIME DURING THE CALENDAR YEAR

[ ][ ] (enter code)

5. RELEASES OF THE CHEMICAL TO THE ENVIRONMENT ON-SITE

You may report releases of less than 1,000 pounds by checking ranges under A.1. (Do not use both A.1 and A.2)		A. Total Release (pounds/year)		B. Basis of Estimate	C. % From Stormwater
		A.1 Reporting Ranges 0 1–499 500–999	A.2 Enter Estimate	(enter code)	
5.1 Fugitive or non-point air emissions	5.1a	[ ] [ ] [ ]		5.1b [ ]	
5.2 Stack or point air emissions	5.2a	[ ] [ ] [ ]		5.2b [ ]	
5.3 Discharges to receiving streams or water bodies 5.3.1 [ ]	5.3.1a	[ ] [ ] [ ]		5.3.1b [ ]	5.3.1c %
(Enter letter code for stream from Part I Section 3.10 in the box provided.) 5.3.2 [ ]	5.3.2a	[ ] [ ] [ ]		5.3.2b [ ]	5.3.2c %
5.3.3 [ ]	5.3.3a	[ ] [ ] [ ]		5.3.3b [ ]	5.3.3c %
5.4 Underground injection on-site	5.4a	[ ] [ ] [ ]		5.4b [ ]	
5.5 Releases to land on-site 5.5.1 Landfill	5.5.1a	[ ] [ ] [ ]		5.5.1b [ ]	
5.5.2 Land treatment/application farming	5.5.2a	[ ] [ ] [ ]		5.5.2b [ ]	
5.5.3 Surface impoundment	5.5.3a	[ ] [ ] [ ]		5.5.3b [ ]	
5.5.4 Other disposal	5.5.4a	[ ] [ ] [ ]		5.5.4b [ ]	

[ ] (Check if additional information is provided on Part IV-Supplemental Information.)

EPA Form 9350-1 (1-90) Revised - Do not use previous versions.

⊕EPA

EPA FORM R

PART III. CHEMICAL-SPECIFIC INFORMATION
(continued)

(This space for your optional use.)

---

6. TRANSFERS OF THE CHEMICAL IN WASTE TO OFF-SITE LOCATIONS

You may report transfers of less than 1,000 pounds by checking ranges under A.1. (Do not use both A.1 and A.2)	A. Total Transfers (pounds/year)		B. Basis of Estimate (enter code)	C. Type of Treatment/ Disposal (enter code)
	A.1 Reporting Ranges 0 1–499 500–999	A.2 Enter Estimate		
6.1.1 Discharge to POTW (enter location number from Part II, Section 1.) [1].	[ ] [ ] [ ]		6.1.1b [ ]	▓▓▓
6.2.1 Other off-site location (enter location number from Part II, Section 2.) [2].	[ ] [ ] [ ]		6.2.1b [ ]	6.2.1c [M][ ][ ]
6.2.2 Other off-site location (enter location number from Part II, Section 2.) [2].	[ ] [ ] [ ]		6.2.2b [ ]	6.2.2c [M][ ][ ]
6.2.3 Other off-site location (enter location number from Part II, Section 2.) [2].	[ ] [ ] [ ]		6.2.3b [ ]	6.2.3c [M][ ][ ]

[ ] (Check if additional information is provided on Part IV–Supplemental Information.)

---

7. WASTE TREATMENT METHODS AND EFFICIENCY

[ ] Not Applicable (NA) – Check if no on-site treatment is applied to any wastestream containing the chemical or chemical category.

A. General Wastestream (enter code)	B. Treatment Method (enter code)	C. Range of Influent Concentration (enter code)	D. Sequential Treatment? (check if applicable)	E. Treatment Efficiency Estimate	F. Based on Operating Data? Yes No
7.1a [ ]	7.1b [ ][ ][ ]	7.1c [ ]	7.1d [ ]	7.1e ___%	7.1f [ ] [ ]
7.2a [ ]	7.2b [ ][ ][ ]	7.2c [ ]	7.2d [ ]	7.2e ___%	7.2f [ ] [ ]
7.3a [ ]	7.3b [ ][ ][ ]	7.3c [ ]	7.3d [ ]	7.3e ___%	7.3f [ ] [ ]
7.4a [ ]	7.4b [ ][ ][ ]	7.4c [ ]	7.4d [ ]	7.4e ___%	7.4f [ ] [ ]
7.5a [ ]	7.5b [ ][ ][ ]	7.5c [ ]	7.5d [ ]	7.5e ___%	7.5f [ ] [ ]
7.6a [ ]	7.6b [ ][ ][ ]	7.6c [ ]	7.6d [ ]	7.6e ___%	7.6f [ ] [ ]
7.7a [ ]	7.7b [ ][ ][ ]	7.7c [ ]	7.7d [ ]	7.7e ___%	7.7f [ ] [ ]
7.8a [ ]	7.8b [ ][ ][ ]	7.8c [ ]	7.8d [ ]	7.8e ___%	7.8f [ ] [ ]
7.9a [ ]	7.9b [ ][ ][ ]	7.9c [ ]	7.9d [ ]	7.9e ___%	7.9f [ ] [ ]
7.10a [ ]	7.10b [ ][ ][ ]	7.10c [ ]	7.10d [ ]	7.10e ___%	7.10f [ ] [ ]

[ ] (Check if additional information is provided on Part IV–Supplemental Information.)

---

8. POLLUTION PREVENTION: OPTIONAL INFORMATION ON WASTE MINIMIZATION
(Indicate actions taken to reduce the amount of the chemical being released from the facility. See the instructions for coded items and an explanation of what information to include.)

A. Type of Modification (enter code)	B. Quantity of the Chemical in Wastes Prior to Treatment or Disposal			C. Index	D. Reason for Action (enter code)
	Current reporting year (pounds/year)	Prior year (pounds/year)	Or percent change (Check (+) or (−))		
[M]	_____	_____	[ ] + [ ] − ____%	[ ].[ ]	[R][ ]

EPA Form 9350-1 (1-90) Revised – Do not use previous versions.

&EPA

EPA FORM R
PART IV. SUPPLEMENTAL INFORMATION
Use this section if you need additional space for answers to questions in Part III.
Number the lines used sequentially from lines in prior sections (e.g., 5.3.4, 6.1.2, 7.11)

(This space for your optional use.)

ADDITIONAL INFORMATION ON RELEASES OF THE CHEMICAL TO THE ENVIRONMENT ON-SITE (Part III, Section 5.3)

You may report releases of less than 1,000 pounds by checking ranges under A.1. (Do not use both A.1 and A.2)	A. Total Release (pounds/year)		B. Basis of Estimate (enter code in box provided)	C.% From Stormwater
	A.1 Reporting Ranges 0 1–499 500–999	A.2 Enter Estimate		
5.3 Discharges to receiving streams or water bodies 5.3.___ ☐	5.3.___ a [ ] [ ] [ ]		5.3.___ b ☐	5.3.___ c ____ %
(Enter letter code for stream from Part I Section 3.10 in the box provided.) 5.3.___ ☐	5.3.___ a [ ] [ ] [ ]		5.3.___ b ☐	5.3.___ c ____ %
5.3.___ ☐	5.3.___ a [ ] [ ] [ ]		5.3.___ b ☐	5.3.___ c ____ %

ADDITIONAL INFORMATION ON TRANSFERS OF THE CHEMICAL IN WASTE TO OFF-SITE LOCATIONS (Part III, Section 6)

You may report transfers of less than 1,000 pounds by checking ranges under A.1. (Do not use both A.1 and A.2)	A.Total Transfers (pounds/year)		B. Basis of Estimate (enter code in box provided)	C. Type of Treatment/ Disposal (enter code in box provided)
	A.1 Reporting Ranges 0 1–499 500–999	A.2 Enter Estimate		
6.1.___ Discharge to POTW (enter location number from Part II, Section 1.) [1].☐	[ ] [ ] [ ]		6.1.___ b ☐	▓▓▓▓▓
6.2.___ Other off-site location (enter location number from Part II, Section 2.) [2].☐	[ ] [ ] [ ]		6.2.___ b ☐	6.2.___ c [M][ ]
6.2.___ Other off-site location (enter location number from Part II, Section 2.) [2].☐	[ ] [ ] [ ]		6.2.___ b ☐	6.2.___ c [M][ ]
6.2.___ Other off-site location (enter location number from Part II, Section 2.) [2].☐	[ ] [ ] [ ]		6.2.___ b ☐	6.2.___ c [M][ ]

ADDITIONAL INFORMATION ON WASTE TREATMENT METHODS AND EFFICIENCY (Part III, Section 7)

A. General Wastestream (enter code in box provided)	B. Treatment Method (enter code in box provided)	C. Range of Influent Concentration (enter code)	D. Sequential Treatment? (check if applicable)	E. Treatment Efficiency Estimate	F. Based on Operating Data? Yes No
7.___ a ☐	7.___ b ☐☐☐	7.___ c ☐	7.___ d [ ]	7.___ e ____ %	7.___ f [ ] [ ]
7.___ a ☐	7.___ b ☐☐☐	7.___ c ☐	7.___ d [ ]	7.___ e ____ %	7.___ f [ ] [ ]
7.___ a ☐	7.___ b ☐☐☐	7.___ c ☐	7.___ d [ ]	7.___ e ____ %	7.___ f [ ] [ ]
7.___ a ☐	7.___ b ☐☐☐	7.___ c ☐	7.___ d [ ]	7.___ e ____ %	7.___ f [ ] [ ]
7.___ a ☐	7.___ b ☐☐☐	7.___ c ☐	7.___ d [ ]	7.___ e ____ %	7.___ f [ ] [ ]
7.___ a ☐	7.___ b ☐☐☐	7.___ c ☐	7.___ d [ ]	7.___ e ____ %	7.___ f [ ] [ ]
7.___ a ☐	7.___ b ☐☐☐	7.___ c ☐	7.___ d [ ]	7.___ e ____ %	7.___ f [ ] [ ]
7.___ a ☐	7.___ b ☐☐☐	7.___ c ☐	7.___ d [ ]	7.___ e ____ %	7.___ f [ ] [ ]
7.___ a ☐	7.___ b ☐☐☐	7.___ c ☐	7.___ d [ ]	7.___ e ____ %	7.___ f [ ] [ ]

The database is organized so you can obtain data categorized by company, chemical, medium to which the chemical is released, waste treatment, and off-site transfer. Thus, the system can enable you to answer such questions as: How much of chemical X did company Y release at facilities throughout the nation last year? What are the names and addresses of steel plants importing chemical X in city Y? What quantity of chemical X did company Y ship off-site last year? What waste minimization methods are reported by companies in X county?

You can get additional information about the TRI file from:

> TRI Representative
> Specialized Information Services
> National Library of Medicine
> 8600 Rockville Pike
> Bethesda, MD 20894
> Phone: (301) 496-6531

If you do not have access to a computer, the TRI data are available in a public reading room in EPA headquarters. You can find out about this from:

> EPA
> P.O. Box 70266
> Washington, DC 20024-0266
> Phone: (202) 488-1501

In addition, your state contact (see Appendix B) can tell you how to obtain the information at a state location. Finally, the company you are interested in may be willing to give you information, although no company is required by law to give you its TRI forms. Many libraries are also beginning to carry copies of the TRI data.

How To Use TRI Information

The Toxics Release Inventory database gives you access to information about many chemicals at many plants. Thus, it enables you to compare plants and to use this comparison to decide which plants and chemicals you would like to focus on. Several activists have found it useful to organize TRI data for local companies in their own databases. Creating a database may make it easier for you to see patterns of emissions or areas of particular concern. It will also provide you with a ready reference when meeting with industrial representatives.

Once you have the data organized, you will want to identify target chemicals for reduction at individual companies. You can use several different criteria for targeting.

Amount of chemical released to the environment annually. This calculation may involve compiling data on releases of the chemical from all area companies. If several companies are releasing the same chemical, there may be a cumulative health and environment risk that none of the companies has considered.

Degree of threat to public health and environment posed by the chemical. Appendix C gives an overview of the known health effects of each of the TRI chemicals. Upon request, the EPA Right to Know Hotline will provide hazardous substance fact sheets for individual TRI chemicals (see Resources). Since the chemicals on the TRI list vary widely in the degree of hazard they pose, it is useful to sort out potential major problems from minor ones.

Comparison with other companies that produce the same product. Using TOXNET, you can compare the emissions of any company you are interested in to those of all other United States companies in the same industrial category. If this company is producing a lot more pollution than other companies of its type, you can bring this to the attention of the company representatives you meet with and ask why others seem to be able to operate more efficiently. Although there may be unavoidable differences in efficiency due to genuine product differences, it is also possible that other companies have identified more efficient methods of operation. In the latter case, you will be doing the company a favor by bringing the discrepancy to its attention. In either case, as someone who must live with the pollution, you deserve an explanation.

Identification of chemicals that are being treated inefficiently. You can use the TRI treatment efficiency data to single out chemicals that are resistant to treatment; these are excellent targets for source reduction.

Identification of rivers and streams receiving excessive levels of chemicals. By studying TRI data for all area industries, you may find that certain bodies of surface water are receiving large amounts of pollutants. You can bring this information to the attention of all companies involved, or to the state environmental protection agency, and seek reductions.

Waste minimization activities. If the company has not provided the voluntary waste minimization information, you can ask why. If the company has indicated "other" rather than a specific type of waste minimization, you can find out exactly what was done.

Comparison of current year's chemical emissions with the previous year's emissions. Once data are available for 1988 and later years, as well as 1987, you will be able to figure out whether pollution has increased or decreased for any chemical the company reports under TRI. Overall increases in pollution are clearly a concern and should be explained by the company. Did the levels of production increase? Has a new product line been added? Overall decreases in pollution levels are obviously a positive sign, although they may not indicate source reduction. Increased on-site treatment and recycling will decrease emissions levels just as source reduction will. Getting an overview of pollution trends at the company will be helpful even though it will lead to further questions.

What Information TRI Does Not Provide

It is important to understand that the Toxics Release Inventory does not provide information about all pollutants even for facilities that report to TRI.

- There are many potentially dangerous pollutants that are not on the TRI list. Thus, TRI does not necessarily alert you to all the pollutants of concern released by a particular company.

- Even if a chemical is on the TRI list, it does not have to be reported if it is released in quantities smaller than the specified thresholds. These releases may still be of concern if they create high exposure levels, through releases into drinking water by a lot of different companies, for example.

- Companies do not have to report the quantity of pollutant they are treating on-site or recycling. Therefore, TRI does not provide a total accounting of pollutants created at a given facility.

Resources

OMB Watch. *Community Right-to-Know: A New Tool for Pollution Prevention.* (Available from OMB Watch, 2001 O Street NW, Washington DC 20036. (202) 659-1711.)

OMB Watch. *Using Community Right to Know: A Guide to a New Federal Law.* (Available from OMB Watch, 2001 O Street NW, Washington DC 20036. (202) 659-1711.)

Toxics Coordinating Project (Ted Smith). *Citizen's Guide to the New Federal Right-to-Know Law: How you can get toxics information and use it to fight toxic pollution.* (Available from Toxics Coordinating Project, 942 Market Street, #502, San Francisco, CA 94102. (415) 781-2745.)

United States Environmental Protection Agency Office of Toxic Substances. *A Guide to the Toxic Chemical Release Inventory Form R.* EPA 560/4-88-010, September 1988. (Available through the Emergency Planning and Community Right to Know Information Hotline. Toll-free (800) 535-0202; in Washington DC (202) 479-2449; 8:30 AM - 7:30 PM Eastern time.)

United States Environmental Protection Agency Office of Toxic Substances. *Common Synonyms for Chemicals Listed under Section 313 of the Emergency Planning and Community Right to Know Act.* January 1988. (Available from the EPA Office of Toxic Substances, Washington DC 20460.)

Provides other names you may encounter for chemicals listed in the Toxics Release Inventory, thus enabling you to look them up in the database.

The Toxics Release Inventory

With the advent of this inventory,
industry becomes accountable to the public as never before

By David Sarokin

Among the provisions of the Emergency Planning and Community Right to Know Law of 1986 is a requirement for EPA to establish and maintain a national toxic chemical inventory in a computer data base. The data base has come to be known as the Toxics Release Inventory (TRI). Although only in its first year of implementation, TRI has already been referred to as a "revolution" in the way government, industry, and the public deal with the risks posed by toxic chemicals.

TRI contains basic information on toxic chemicals and their manufacturers, including data on chemical identity, quantity of environmental release, and types of waste treatment, as well as an opportunity for companies to quantify their achievements in reducing waste generation. Information for this data base, which is updated annually, is supplied by inventory forms completed by the users of toxic chemicals in the manufacturing sector.

Does EPA need yet another information collection form on toxic chemicals? Absolutely. For a variety of reasons, decades of toxic chemical regulation, management, and data collection have not yielded a very useful picture of where toxic chemicals are, or which chemicals and what quantities are entering the environment.

For example, an agency report on toxic chemical air emissions across the nation concluded that the absence of reliable emissions data was the major stumbling block in assessing the scope of the air toxics problem. Similar conclusions have been expressed regarding wastewater discharges to streams, releases to sewage treatment plants, toxic chemical constituents of RCRA wastes, and deep-well injection. The massive quantities of data that have been collected under EPA's major environmental programs were intended primarily as tools for regulatory oversight, and they are quite effective in this regard. But when EPA turned to the same data

David Sarokin

to address fundamental questions of where, what, and how much—questions these regulatory systems were not designed to answer—the limitations of the data were evident.

With the advent of a nationwide toxics release inventory, EPA will have, for the first time, a uniform and fairly comprehensive picture of toxic chemical releases to the environment and waste management practices. Because the data are collected annually, TRI also will provide a clear picture of trends pertaining to changing patterns of toxic chemical wastes.

In other words, EPA's multifaceted environmental control programs have suffered something of a "Tower of Babel" syndrome, with many different regulatory tongues speaking in the mutually incomprehensible languages of toxic chemicals. TRI will provide a commonality of data, which will permit a much more comprehensive picture than has been possible in the past, despite the overwhelming quantities of data already collected.

One other feature of TRI that stands out is the one most in keeping with the community-right-to-know focus of the law that created it: TRI is designed for public access. For example, anyone with a desire to learn about toxic chemicals at the industrial plant across the street will have easy access to that plant's TRI submission. Anyone with a

computer, modem, and a modicum of skill at data manipulation can phone into TRI and access information on a given town, state, industry, or chemical for detailed analysis. For the first time, citizens will be able to ask, How many tons of toxic chemicals are emitted in my community? By whom? and Where?

TRI reporting

TRI regulations list 308 chemicals and 20 chemical categories that are subject to reporting requirements. But not everyone who handles these chemicals is required to submit a TRI report. A facility must report if it is classified as a manufacturer (i.e., it has a Standard Industrial Classification code of 20–39); it has 10 or more full-time employees; and it handles TRI-listed chemicals above threshold amounts.

Any facility that meets all three criteria is subject to TRI requirements. The threshold amounts that trigger reporting are 75,000 lb/year for chemicals that are manufactured, imported, or processed, and 10,000 lb/year for chemicals used in any other manner. Although the 75,000 figure is the threshold for reporting 1987 data, the requirement will drop to 50,000 lb/year for 1988 reports and 25,000 lb/year in subsequent years.

To complete the reporting form, manufacturers must include a page of information on facility identification and a page identifying all off-site locations to which toxic chemical wastes are transferred. In addition, for each reported chemical, manufacturers must include two pages of detailed information on chemical uses in manufacturing and processing; releases (in pounds per year) to air, including point source and

Hot line. EPA has a hot line for information and assistance on the Toxics Release Inventory: 1-800-535-0202. In Washington, D.C., and Alaska call (202) 479-2449.

fugitive emissions, waterways, underground injection, and land; transfers to off-site waste treatment or disposal facilities; types and efficiencies of on-site treatment processes used; and optional information on waste minimization.

Opportunities to declare information a trade secret are fairly limited under TRI. Only the chemical identity can be claimed as proprietary, and only after submitting a detailed justification for the claim; there are penalties for submitting frivolous claims. When a chemical's identity is kept as a trade secret, generic information on the health and environmental effects of the chemical must be made available in its stead. The reports for calendar year 1987 are due to EPA and to states by July 1, 1988.

Estimating releases

Companies are not required to undertake any monitoring or measurements for TRI beyond those already in use for other environmental reporting. Data already on hand for air and water permits and RCRA record-keeping can be used for compiling TRI data such as direct measurements, calculated releases, or estimates. Nevertheless, gathering or calculating the data for TRI reporting will doubtlessly be a demanding task for many companies, particularly when chemicals have multiple uses at a single facility and can find their way into dozens of discrete waste streams.

EPA has prepared a guidance manual to assist companies in using available data to calculate or estimate air, water, and solid waste generation and treatment efficiencies. In addition, the agency is preparing guidance documents for specific industries and chemical uses. The documents cover four basic categories of release estimates using monitoring data, mass balance approaches, published emission factors, engineering calculations, or judgments.

Data quality promises to be a chief concern as TRI numbers begin to surface. The EPA Office of Toxic Substances, which manages the TRI program, is planning a multipronged approach to data quality maintenance, including training on TRI estimations for both governmental and industrial personnel, computerized data checks to trigger a red flag for questionable data, telephone and field follow-up with facilities to audit data quality, and a vigorous program of compliance and enforcement.

Uses of the inventory data

EPA anticipates receiving reports from as many as 30,000 facilities nationwide, with an average of 10 reportable chemicals, for a total of 300,000 reports. These reports will represent a detailed numerical picture of toxic chemicals in the manufacturing sector. The data base will be made fully accessible to the public by the spring of 1989, although some TRI information on individual facilities may be accessible before this date.

Like any versatile information system, the uses of TRI data will be limited only by the needs and imagination of the system users, who include members of the general public, industry, academia, environmental groups, the news media, and all levels of government active in environmental issues. Although there is no way of foreseeing all the uses to which the data will be put, several types of requests and analyses are expected to be fairly routine.

Facility requests. Citizens and community groups will request release data on individual facilities in their neighborhoods in order to focus their concerns about toxic chemicals. Government officials will use TRI data as part of their routine information gathering and oversight activities for individual facilities.

Geographic requests. Summary reports of data for states, counties, EPA regions, and other geographic areas are likely to be common. Reporting formats could include information such as total quantities of air, water, and solid wastes as well as more detailed information on the largest sources of toxic wastes and the largest volume chemical releases reported.

Chemical requests. Reports of individual chemical releases or categories of chemicals such as chlorinated hydrocarbons will be requested.

Industry requests. These reports will include comparisons of facilities within a given industrial sector in terms of chemicals used and released, treatment methods and efficiencies, and the extent and impact of waste reduction practices.

The system will have the flexibility to process hierarchical requests for, say, total releases of benzene from chemical plants in a given state or region.

The impact of the inventory

There are costs to industry for providing TRI data, but EPA and state governments, which receive the TRI forms, also incur costs for managing and analyzing the information—and these costs are by no means trivial. The analysis by the Office of Toxic Substances shows that TRI regulations will require an industrial company to spend, on average, more than 400 hours filling out TRI data for 10 chemicals. In addition, the 30,000 reporting facilities will spend a total of $500 million.

A sizable expenditure, to be sure, but it is also an important investment toward better and more efficient management of toxic chemical wastes—an investment from which all interested parties stand to benefit. Among the beneficiaries are the public, industry,

---

Filling out forms on releases

An expert system has been developed to aid engineers in the complex process of completing the Toxic Chemical Resease Inventory (Form R). This computer software package, called the 313 Advisor, was developed by Du Pont Environmental Management Services. For information, call (800) 532-7233.

This material is not endorsed by the author nor by EPA.

TRI is distinguished from other environmental data

Multimedia. TRI collects data on releases to all environmental media—air, water, and land—in contrast to traditional regulatory programs that focus on only a single environmental media.

Program independent. TRI is not constrained by the inclusions, categorizations, loopholes, or exemptions of existing regulatory programs and can therefore create a much more comprehensive picture of toxic chemicals than has been possible in the past. Whether a chemical is or is not a new existing source of hazardous air pollutant (NESHAP) air pollution, a national pollution discharge elimination system (NPDES) priority pollutant, or a listed Resource Conservation and Recovery Act (RCRA) waste is irrelevant to TRI; facilities that are subject to TRI reporting must account for all releases of the more than 300 listed chemicals.

Chemical specific. By collecting data on individual chemicals in 20 chemical categories, TRI avoids the ambiguities inherent in traditional schemes that aggregate dozens of chemicals under a single reporting classification such as chemical oxygen demand, volatile organic chemicals, or F002 hazardous wastes. TRI data can be linked to toxicological data that also tend to be chemical specific.

Data base design. TRI is the first major environmental program intended from the outset to serve as a data source on toxic chemical releases. Attempts to retrofit air, water, and RCRA data into a computerized data base have met with limited success because of the variability in data formats and state-by-state data management. The uniformity of TRI may well be its most potent asset.

and the government.

The public. TRI, along with other provisions of the Emergency Planning and Community Right to Know Act, is designed to empower the public to learn about and become involved in environmental management decisions that pertain to toxic chemicals in their own communities. The relatively straightforward nature of the data, reported in pounds per year of chemical release, and the ease of access will be a step toward meeting the congressional goal of insuring the public's right to know.

Industry. Increased industry awareness at upper management levels—both of the quantity of materials lost as wastes and of the costs and potential liability of these losses—is often an effective spur to reducing loss and thus safeguarding the environment and the financial bottom line at the same time. TRI will be an important tool for fostering waste reduction. In addition, the release data—particularly the data on waste minimization—will provide facilities that practice waste reduction an opportunity to document their progress in reducing toxic chemical releases.

Government. Which environmental programs are working well? Which are not? What chemicals are effectively controlled, and which escape through the regulatory net? TRI will enable EPA, states, and others to answer these questions and to commit program resources to achieving the best environmental results for a finite amount of resources. The documentation of trends in toxic chemical waste releases and management will provide unambiguous evidence of the impact of regulatory controls.

The question of risks

What will happen when the public learns that a certain amount of a given chemical, perhaps a carcinogen, is being released as waste to the air, water, or soil by Company X down the road? Release data alone is a poor indicator of environmental concentrations or ultimate exposure. Chemical and environmental properties complicate our understanding of risk because context plays a large role. For example, releases of 87,000 lb/year sounds much worse to some than 10 pounds per hour, even though the two are equivalent.

Even when adequate health and fate data are available, formal risk assessments are costly and cumbersome; in the end, the assessment is difficult to describe meaningfully to a concerned community. As TRI reveals a large number of apparent toxics problems, the situation will become even more complicated; sizable emission numbers do not necessarily mean sizable risk, but these numbers can generate substantial community concern just the same.

The onus of giving meaning to TRI numbers will fall on the public, government, and industry. Companies likely will find themselves justifying the release of a particular chemical or demonstrating to the public their efforts to reduce releases. States, EPA, and other environmental agencies will need to reevaluate the efficiency of their own control programs in light of legitimate questions from the public. The public, for its part, will need to understand that waste is an inevitable part of industrial processes and that zero risk can at best be a goal to strive for, but not one ever to be attained.

David Sarokin is an environmental protection specialist with EPA's Office of Toxic Substances. Previously, Sarokin worked on industrial waste reduction issues for the state of New Jersey. He is also the author of Cutting Chemical Wastes, *published by Inform, Inc.*

This article originally appeared in *Environmental Science & Technology,* June 1988.

Appendix B

State Contacts for TRI Information

Alabama

Mr. E. John Williford, Chief of Operations
Alabama Emergency Response Commission
Alabama Department of Environmental
  Management
1751 Congressman W.L. Dickinson Drive
Montgomery, AL 36109
(205) 271-7700

Alaska

Mr. Dennis Kelso, Chair
Alaska State Emergency Response
  Commission
P.O. Box O
Juneau, AK 99811
(907) 465-2600

American Samoa

Mr. Pati Faiai, Director
American Samoa EPA
Office of the Governor
Pago Pago, AS 96799
International Number (684) 633-2304

Arizona

Mr. Carl F. Funk, Executive Director
Arizona Emergency Response Commission
Division of Emergency Services
5636 East McDowell Road
Phoenix, AZ 85008
(602) 231-6326

Arkansas

Ms. Becky Bryant
Depository of Documents
Arkansas Department of Labor
10421 West Markham
Little Rock, AR 72205
(501) 682-4534

California

Mr. Chuck Shulock
Office of Environmental Affairs
P.O. Box 2815
Sacramento, CA 95812
Attn: Section 313 Reports
(916) 324-8124
(916) 322-7236 Completed Form R
  Information

Colorado

Colorado Emergency Planning Commission
Colorado Department of Health
4210 East 11th Avenue
Denver, CO 80220
Attn: Judy Waddill
(303) 331-4858

Commonwealth of Northern Mariana Islands

Mr. Russell Meecham, III
Division of Environmental Quality
P.O. Box 1304
Saipan, CNMI 96950
(670) 234-6984

Connecticut

Ms. Sue Vaughn, Title III Coordinator
State Emergency Response Commission
Department of Environmental Protection
State Office Building, Room 161
165 Capitol Avenue
Hartford, CT 06106
(203) 566-4856

Delaware

Mr. Robert French
Chief Program Administrator
Air Resource Section
Department of Natural Resources and
  Environmental Control
P.O. Box 1401
Dover, DE 19903
(302) 736-4791

District of Columbia

Mr. Joseph P. Yeldell, Chairman
District of Columbia Emergency Response
  Commission
Office of Emergency Preparedness
2000 14th Street, NW
Frank Reeves Center for Municipal Affairs
Washington, DC 20009
(202) 727-6161

Florida

Mr. Thomas G. Pelham, Chairman
Florida Emergency Response Commission
Secretary, Florida Department of Community
  Affairs
2740 Centerview Drive
Tallahassee, FL 32399-2149
(904) 488-1472
In Florida: (800) 635-7179

Georgia

Mr. Jimmy Kirkland
Georgia Emergency Response Commission
205 Butler Street, SE
Floyd Tower East
11th Floor, Suite 1166
Atlanta, GA 30334
(404) 656-6905

Guam

Mr. Roland Solidio
Guam EPA
P.O. Box 2999
Aguana, GU 96910
(671) 646-8863

Hawaii

Mr. John C. Lewin, M.D., Chairman
Hawaii State Emergency Response
  Commission
Hawaii State Department of Health
P.O. Box 3378
Honolulu, HI 96801-9904
(808) 548-6505

Idaho

Idaho Emergency Response Commission
State House
Boise, ID 83720
Attn: Ms. Jenny Records
(208) 334-5888

Illinois

Mr. Joe Goodner
Emergency Planning Unit
Illinois EPA
P.O. Box 19276
2200 Churchill Road
Springfield, IL 62794-9276
(217) 782-3637

Indiana

Mr. Philip Powers, Director
Indiana Emergency Response Commission
5500 West Bradbury Avenue
Indianapolis, IN 46241
(317) 243-5176

Iowa

Department of Natural Resources
Records Department
900 East Grand Avenue
Des Moines, IA 50319
(515) 281-8852

Kansas

Right-to-Know Program
Kansas Department of Health and
  Environment
Mills Building, 5th Floor
109 S.W. 9th Street
Topeka, KS 66612
(913) 296-1690

Kentucky

Ms. Valerie Hudson
Kentucky Department of Environmental
  Protection
18 Reilly Road
Frankfort, KY 40601
(502) 564-2150

Louisiana

Mr. R. Bruce Hammatt
Emergency Response Coordinator
Department of Environmental Quality
P.O. Box 44066
333 Laurel Street
Baton Rouge, LA 70804-4066
(504) 342-8617

Maine

Mr. David D. Brown, Chairman
State Emergency Response Commission
Station Number 72
Augusta, ME 04333
(207) 289-4080
In Maine: (800) 452-8735

Maryland

Ms. Marsha Ways
State Emergency Response Commission
Maryland Department of the Environment
Toxics Information Center
2500 Broening Highway
Baltimore, MD 21224
(301) 631-3800

Massachusetts

Mr. Arnold Sapenter
c/o Title III Emergency Response
  Commission
Department of Environmental Quality
  Engineering
One Winter Street, 10th Floor
Boston, MA 02108
(617) 292-5993

Michigan

Title III Coordinator
Michigan Department of Natural Resources
Environmental Response Division
Title III Notification
P.O. Box 30028
Lansing, MI 48909
(517) 373-8481

Minnesota

Mr. Lee Tishler, Director
Minnesota Emergency Response
  Commission
290 Bigelow Building
450 North Syndicate
St. Paul, MN 55155
(612) 643-3000

Mississippi

Mr. J.E. Maher, Chairman
Mississippi Emergency Response
  Commission
Mississippi Emergency Management Agency
P.O. Box 4501
Fondren Station
Jackson, MS 39296-4501
(601) 960-9973

Missouri

Mr. Dean Martin, Coordinator
Missouri Emergency Response Commission
Missouri Department of Natural Resources
P.O. Box 3133
Jefferson City, MO 65102
(314) 751-7929

Montana

Mr. Tom Ellerhoff, Co-Chairman
Montana Emergency Response Commission
Environmental Sciences Division
Department of Health & Environmental
  Sciences
Cogswell Building A-107
Helena, MT 59620
(406) 444-6911

Nebraska

Mr. Clark Smith, Coordinator
Nebraska Emergency Response
  Commission
Nebraska Department of Environmental
  Control
P.O. Box 98922
State House Station
Lincoln, NE 68509-8922
(402) 471-2186

Nevada

Mr. Bob King, Director
Division of Emergency Management
2525 South Carson Street
Carson City, NV 89710
(702) 885-4240

New Hampshire

Mr. George L. Iverson, Director
State Emergency Management Agency
Title III Program
State Office Park South
107 Pleasant Street
Concord, NH 03301
(603) 271-2231

New Jersey

New Jersey Emergency Response
  Commission
SARA Title III Section 313
Department of Environmental Protection
Division of Environmental Quality
Bureau of Hazardous Substances
  Information
CN-405
Trenton, NJ 08625
(609) 292-6714

New Mexico

Mr. Samuel Larcombe
New Mexico Emergency Response
  Commission
New Mexico Department of Public Safety
P.O. Box 1628
Santa Fe, NM 87504-1628
(505) 827-9222

New York

New York Emergency Response
  Commission
New York State Department of
  Environmental Conservation
Bureau of Spill Response
50 Wolf Road/Room 326
Albany, NY 12233-3510
(518) 457-4107

North Carolina

North Carolina Emergency Response
  Commission
North Carolina Division of Emergency
  Management
116 West Jones Street
Raleigh, NC 27603-1335
(919) 733-3867

North Dakota

SARA Title III Coordinator
North Dakota State Department of Health and
  Consolidated Laboratories
1200 Missouri Avenue
P.O. Box 5520
Bismarck, ND 58502-5520
(701) 224-2374

Ohio

Ms. Cindy Sferra-DeWulf
Division of Air Pollution Control
1800 Watermark Drive
Columbus, OH 43215
(614) 644-2266

Oklahoma

Emergency Response Commission
Office of Civil Defense
P.O. Box 53365
Oklahoma City, OK 73152
(405) 521-2481

Oregon

Mr. Ralph M. Rodia
Oregon Emergency Response Commission
c/o State Fire Marshall
3000 Market Street Plaza, Suite 534
Salem, OR 97310
(503) 378-2885

Pennsylvania

Mr. James Tinney
Bureau of Right-to-Know
Room 1503
Labor and Industry Building
7th & Forrester Streets
Harrisburg, PA 17120
(717) 783-2071

Puerto Rico

SERC Commissioner
Title III-SARA Section 313
Puerto Rico Environmental Quality Board
P.O. Box 11488
Santurce, PR 00910
(809) 722-0077

Rhode Island

Department of Environmental Management
Division of Air and Hazardous Materials
291 Promenade Street
Providence, RI 02908
Attn: Toxic Release Inventory
(401) 277-2808

South Carolina

Mr. Ron Kinney
Department of Health and Environmental
  Control
2600 Bull Street
Columbia, SC 29201
(803) 734-5200

South Dakota

Ms. Lee Ann Smith, Director
South Dakota Emergency Response
  Commission
Department of Water and Natural Resources
Joe Foss Building
523 East Capitol
Pierre, SD 57501-3181
(605) 773-3153

Tennessee

Mr. Lacy Suiter, Chairman
Tennessee Emergency Response
  Commission
Director, Tennessee Emergency
  Management Agency
3041 Sidco Drive
Nashville, TN 37204
(615) 252-3300
(800) 262-3300 (in Tennessee)
(800) 258-3300 (out of state)

Texas

Mr. David Barker, Supervisor
Emergency Response Unit
Texas Water Commission
P.O. Box 13087—Capitol Station
Austin, TX 78711-3087
(512) 463-8527

Utah

Mr. Neil Taylor
Utah Hazardous Chemical Emergency
  Response Commission
Utah Division of Environmental Health
288 North 1460 West
P.O. Box 16690
Salt Lake City, UT 84116-0690
(801) 538-6121

Vermont

Dr. Jan Carney, Commissioner
Department of Health
60 Main Street
P.O. Box 70
Burlington, VT 05402
(802) 863-7281

Virginia

Mr. Wayne Halbleib, Director
Virginia Emergency Response Council
Department of Waste Management
James Monroe Building, 14th Floor
101 North 14th Street
Richmond, VA 23219
(804) 225-2513

Virgin Islands

Mr. Allan D. Smith, Commissioner
Department of Planning and Natural
  Resources
U.S. Virgin Islands Emergency Response
  Commission
Title III
Nisky Center, Suite 231
Charlotte Amalie
St. Thomas, VI 00802
(809) 774-3320/Ext. 169 or 170

Washington

Mr. John Ridgway, Chairman
Washington State Department of Ecology
Hazardous Substance Information Office
Mail Stop PV/11
Olympia, WA 98504
(206) 438-7252

West Virginia

Mr. Carl L. Bradford, Director
West Virginia Emergency Response
  Commission
West Virginia Office of Emergency Services
State Capital Building 1, Room EB-80
Charleston, WV 25305
(304) 348-5380

Wisconsin

Department of Natural Resources
P.O. Box 7921
Madison, WI 53707
Attn: Russ Dumst
(608) 266-9255

Wyoming

Mr. Ed Usui, Executive Secretary
Wyoming Emergency Response Commission
Wyoming Emergency Management Agency
Comprehensive Emergency Management
P.O. Box 1709
Cheyenne, WY 82003
(307) 777-7566

Source: These names and addresses are provided by the Environmental Protection Agency in
its *Toxic Chemical Release Inventory Reporting
Package for 1989*, EPA 560/4-90-001, January
1990.

Overview of Known
Health Effects of TRI Chemicals

Table C-1 Alphabetical Listing

Chemical Name	CAS Number	Carcinogenicity	Heritable genetic and chromosomal mutations	Developmental toxicity (including teratogenicity)	Reproductive toxicity	Acute toxicity	Chronic (system) toxicity	Neurotoxicity	Environmental toxicity	Bioaccumulation	Persistence in the Environment
Acetaldehyde	75-07-0	X				X			X		
Acetamide	60-35-5	X									X
Acetone	67-64-1						X		X		
Acetonitrile	75-05-8			X	X	X	X				X
2-Acetylaminofluorene	53-96-3	X									
Acrolein	107-02-8					X	X		X		
Acrylamide	79-06-1		X		X	X	X	X	X		
Acrylic acid	79-10-7						X				
Acrylonitrile	107-13-1	X		X	X	X			X		
Aldrin	309-00-2	X		X	X	X	X		X	X	
Allyl chloride	107-05-1	X		X		X	X	X	X		
alpha-Naphthylamine	134-32-7	X									
Aluminum oxide	1344-28-1						X				
Aluminum (fume or dust)	7429-90-5										
1-Amino-2-methylanthraquinone	82-28-0	X									
4-Aminoazobenzene	60-09-3	X									
4-Aminobiphenyl	92-67-1	X									
Ammonia	7664-41-7					X	X			X	
Ammonium nitrate (solution)	6484-52-2						X				
Ammonium sulfate (solution)	7783-20-2								X		
Aniline	62-53-3	X				X	X		X		
o-Anisidine hydrochloride	134-29-2	X									
o-Anisidine	90-04-0	X					X	X			
p-Anisidine	104-94-9										
Anthracene	120-12-7						X		X	X	
Antimony	7440-36-0				X						
Arsenic	7440-38-2	X									
Asbestos (friable)	1332-21-4	X					X				
Auramine	492-80-8	X									
Barium	7440-39-3										
Benzal chloride	98-87-3	X				X					
Benzamide	55-21-0										

Table C-1 Alphabetical Listing

Chemical Name	CAS Number	Carcinogenicity	Heritable genetic and chromosomal mutations	Developmental toxicity (including teratogenicity)	Reproductive toxicity	Acute toxicity	Chronic (system) toxicity	Neurotoxicity	Environmental toxicity	Bioaccumulation	Persistence in the Environment
Benzene	71-43-2	X		X	X		X		X		
Benzidine	92-87-5	X				X	X				
p-Benzoquinone	106-51-4					X			X		
Benzotrichloride	98-07-7	X					X				
Benzoyl chloride	98-88-4								X		
Benzoyl peroxide	94-36-0										
Benzyl chloride	100-44-7	X		X		X	X	X	X		
Beryllium	7440-41-7	X					X				
Biphenyl	92-52-4			X			X		X		
Bis(2-ethylhexyl) adipate	103-23-1	X	X								
Bromoform	75-25-2						X	X	X		
1,3-Butadiene	106-99-0	X		X	X		X				
1-Butanol	71-36-3						X				
Butyl acrylate	141-32-2								X		
sec-Butyl alcohol	78-92-2										
tert-Butyl alcohol	75-60-5										
Butyl benzyl phthalate	85-68-7						X		X		
1,2-Butylene oxide	106-88-7										
Butyraldehyde	123-72-8								X		
Cadmium	7440-43-9	X		X	X	X	X		X		
Calcium cyanamide	156-62-7					X	X				
Captan	133-06-2	X	X	X	X	X			X		X
Carbaryl	63-25-2			X	X	X	X	X	X	X	X
Carbon disulfide	75-15-0			X	X		X	X	X		X
Carbon tetrachloride	56-23-5	X		X			X		X		X
Carbonyl sulfide	463-58-1						X				
Catechol	120-80-9					X			X		
Chloramben	133-90-4	X									
Chlordane	57-74-9	X		X	X	X	X	X		X	
Chlorine	7782-50-5					X	X		X		
Chlorine dioxide	10049-04-4			X	X						
Chloroacetic acid	79-11-8					X					
2-Chloroacetophenone	532-27-4					X					
Chlorobenzene	108-90-7						X		X		
Chlorobenzilate	510-15-6	X							X		
Chloroform	67-66-3	X		X	X	X			X		X
Chloromethyl methyl ether	107-30-2	X				X					
Chloroprene	126-99-8		X	X	X		X	X			
Chlorothalonil	1897-45-6	X							X		
Chromium	7440-47-3	X					X		X		
Cobalt	7440-48-4						X				
Copper	7440-50-8			X	X				X		
p-Cresidine	120-71-8	X									
m-Cresol	108-39-4					X			X		
o-Cresol	95-48-7					X			X		
p-Cresol	106-44-5					X			X		
Cresol (mixed isomers)	1319-77-3					X	X		X		
Cumene hydroperoxide	80-15-9					X			X		
Cupferron	135-20-6	X									
Cyanide compounds	57-12-5								X		

Table C-1 Alphabetical Listing

Chemical Name	CAS Number	Carcinogenicity	Heritable genetic and chromosomal mutations	Developmental toxicity (including teratogenicity)	Reproductive toxicity	Acute toxicity	Chronic (system) toxicity	Neurotoxicity	Environmental toxicity	Bioaccumulation	Persistence in the Environment
Cyclohexane	110-82-7								X		
C.I. Acid Blue 9, diammonium salt	2650-18-2										
C.I. Acid Blue 9, disodium salt	3844-45-9	X									
C.I. Acid Green 3	4680-78-8	X									
C.I. Basic Green 4	569-64-2					X			X		
C.I. Basic Red 1	989-38-8	X									
C.I. Direct Black 38	1937-37-7	X									
C.I. Direct Blue 6	2602-46-2	X		X							
C.I. Direct Brown 95	16071-86-6	X									
C.I. Disperse Yellow 3	2832-40-8	X									
C.I. Food Red 15	81-88-9	X					X				
C.I. Food Red 5	3761-53-3	X									
C.I. Solvent Orange 7	3118-97-6										
C.I. Solvent Yellow 14	842-07-9	X									
C.I. Solvent Yellow 3	97-56-3	X								X	
C.I. Vat Yellow 4	128-66-5	X									
2,4-D	94-75-7			X	X	X	X		X		
Decabromodiphenyl oxide	1163-19-5	X									
Diallate	2303-16-4	X			X				X		
2,4-Diaminoanisole	615-05-4	X									
2,4-Diaminoanisole sulfate	39156-41-7	X									
4,4'-Diaminodiphenyl ether	101-80-4	X					X				
2,4-Diaminotoluene	95-80-7	X	X								
Diazomethane	334-88-3										
Dibenzofuran	132-64-9										
1,2-Dibromo-3-chloropropane	96-12-8	X		X	X	X	X		X		
Dibutyl phthalate	84-74-2			X	X		X		X	X	
Dichlorobenzene	25321-22-6	X					X		X	X	
m-Dichlorobenzene	541-73-1								X		
o-Dichlorobenzene	95-50-1								X	X	
p-Dichlorobenzene	106-46-7	X					X		X	X	
3,3'-Dichlorobenzidine	91-94-1	X									
Dichlorobromomethane	75-27-4										X
Dichloroethyl ether	111-44-4	X				X	X				
1,1-Dichloroethylene	75-35-4	X		X	X		X		X		
Dichloroisopropyl ether	108-60-1	X				X	X				
Dichloromethyl ether	542-88-1	X					X				
2,4-Dichlorophenol	120-83-2								X		
1,3-Dichloropropene	542-75-6	X				X	X		X		
Dichlorvos	62-73-7			X	X	X			X		X
Dicofol	115-32-2	X				X			X		
1:2,3:4-Diepoxybutane	1464-53-5	X	X			X					
Diethanolamine	111-42-2	X							X		
Diethyl sulfate	64-67-5	X	X								
Diethylhexyl phthalate	117-81-7	X	X	X	X		X		X	X	
Diethylphthalate	84-66-2								X		
3,3'-Dimethoxybenzidine	119-90-4	X									
Dimethyl sulfate	77-78-1	X	X			X			X		
4-Dimethylaminoazobenzene	60-11-7	X				X	X			X	
N,N-Dimethylaniline	121-69-7						X	X			

Table C-1 Alphabetical Listing

Chemical Name	CAS Number	Carcinogenicity	Heritable genetic and chromosomal mutations	Developmental toxicity (including teratogenicity)	Reproductive toxicity	Acute toxicity	Chronic (system) toxicity	Neurotoxicity	Environmental toxicity	Bioaccumulation	Persistence in the Environment
3,3'-Dimethylbenzidine	119-93-7	X									
Dimethylcarbomyl chloride	79-44-7	X				X					
1,1-Dimethylhydrazine	57-14-7	X				X			X		
2,4-Dimethylphenol	105-67-9								X		
Dimethylphthalate	131-11-3					X			X		
2,4-Dinitrophenol	51-28-5			X	X	X	X		X		
2,4-Dinitrotoluene	121-14-2	X					X		X		
2,6-Dinitrotoluene	606-20-2					X			X		
4,6-Dinitro-o-cresol	534-52-1					X			X		
Dioxane	123-91-1	X									
1,2-Diphenylhydrazine	122-66-7	X							X		
Di-n-octylphthalate	117-84-0					X	X			X	
Di-n-propylnitrosamine	621-64-7	X									
Epichlorohydrin	106-89-8	X	X		X	X	X		X		X
Ethyl acrylate	140-88-5	X		X			X		X		X
Ethyl carbamate (urethane)	51-79-6	X	X								
Ethyl chloride	75-00-3										
Ethyl chloroformate	541-41-3					X					
Ethylbenzene	100-41-4			X	X		X		X		
Ethylene	74-85-1						X				
Ethylene dibromide	106-93-4	X	X	X	X	X	X		X		
Ethylene dichloride	107-06-2	X	X								X
Ethylene glycol	107-21-1						X				
Ethylene glycol monoethyl ether	110-80-5			X	X		X				
Ethylene oxide	75-21-8	X	X	X	X	X	X		X		X
Ethyleneimine	151-56-4	X	X		X	X	X		X		X
Ethylenethiourea	96-45-7	X	X	X	X		X				
Fluometuron	2164-17-2	X					X		X		
Formaldehyde	50-00-0	X	X		X		X		X		
Freon 113	76-13-1						X				
Heptachlor	76-44-8	X				X	X		X	X	
Hexachlorobenzene	118-74-1	X		X	X	X	X		X	X	
Hexachlorobutadiene	87-68-3	X		X	X	X	X		X	X	
Hexachlorocyclopentadiene	77-47-4			X	X	X	X	X	X	X	X
Hexachloroethane	67-72-1	X		X	X		X				
Hexachloronaphthalene	1335-87-1										
Hexamethylphosphoramide	680-31-9	X	X				X				
Hydrazine	302-01-2	X					X		X		
Hydrazine sulfate	10034-93-2	X									
Hydrochloric acid	7647-01-0					X	X				
Hydrogen cyanide	74-90-8					X	X		X		
Hydrogen fluoride	7664-39-3			X	X	X	X				
Hydroquinone	123-31-9					X	X		X		
Isobutyraldehyde	78-84-2						X				
Isopropyl alcohol	67-63-0	X					X	X			
4,4'-Isopropylidenediphenol	80-05-7										
Lead	7439-92-1			X	X			X			
Lindane	58-89-9	X		X	X	X			X	X	
Maleic anhydride	108-31-6										
Maneb	1247-38-2			X	X		X	X	X		

Table C-1 Alphabetical Listing

Chemical Name	CAS Number	Carcinogenicity	Heritable genetic and chromosomal mutations	Developmental toxicity (including teratogenicity)	Reproductive toxicity	Acute toxicity	Chronic (system) toxicity	Neurotoxicity	Environmental toxicity	Bioaccumulation	Persistence in the Environment
Manganese	7439-96-5							X			
MBI	101-68-8										
Melamine	108-78-1	X			X						
Mercury	7439-97-6				X		X	X	X		
Methanol	67-56-1							X			
Methoxychlor	72-43-5			X	X			X	X	X	
2-Methoxyethanol	109-86-4			X	X		X				
Methyl acrylate	96-33-3								X		
Methyl bromide	74-83-9					X	X	X	X		X
Methyl chloride	74-87-3			X	X		X				X
1-Methyl ethyl benzene (Cumene)	98-82-8						X	X	X		
Methyl ethyl ketone (MEK)	78-93-3			X	X		X	X			
Methyl hydrazine	60-34-4	X				X	X		X		
Methyl iodide	74-88-4	X						X			X
Methyl isobutyl ketone	108-10-1						X	X			
Methyl isocyanate	624-83-9					X					
Methyl methacrylate	80-62-6				X		X				
Methyl tert-butyl ether	1634-04-4										
Methylene bromide	74-95-3							X			
Methylene chloride	75-09-2	X									X
4,4'-Methylenebis (2-chloroaniline)	101-14-4	X									
4,4'-Methylenebis (N,N-dimethyl)	101-61-1	X									
4,4'-Methylenedianiline	101-77-9						X				
Michler's ketone	90-94-8	X									
Molybdenum trioxide	1313-27-5					X	X	X			
Mustard gas	505-60-2	X	X			X					
Naphthalene	91-20-3								X	X	
beta-Naphthylamine	91-59-8	X									
Nickel	7440-02-0	X		X	X		X		X		
Nitric acid	7697-37-2					X					
Nitrilotriacetic acid	139-13-9	X					X				
Nitrobenzene	98-95-3				X	X	X		X		
4-Nitrobiphenyl	92-93-3	X									
Nitrofen	1836-75-5	X		X	X		X	X	X		
Nitrogen mustard	51-75-2	X	X	X	X	X					
Nitroglycerin	55-63-0								X		
2-Nitrophenol	88-75-5						X	X	X		
p-Nitrophenol	100-02-7					X	X	X	X		
2-Nitropropane	79-46-9	X		X	X	X	X				
N-Nitrosodiethylamine	55-18-5	X	X	X	X	X					
N-Nitrosodimethylamine	62-75-9	X	X			X					
N-Nitrosodiphenylamine	86-30-6	X							X		
p-Nitrosodiphenylamine	156-10-5	X									
N-Nitrosodi-n-butylamine	924-16-3	X									
N-Nitrosomethylvinylamine	4549-40-0	X	X			X					
N-Nitrosomorpholine	59-89-2	X	X	X				X			
N-Nitrosonornicotine	16543-55-8	X									
N-Nitrosopiperidine	100-75-4	X	X			X	X				
N-Nitroso-N-ethyl urea	759-73-9	X	X	X	X						
N-Nitroso-N-methylurea	684-93-5	X	X	X		X					

Table C-1 Alphabetical Listing

Chemical Name	CAS Number	Carcinogenicity	Heritable genetic and chromosomal mutations	Developmental toxicity (including teratogenicity)	Reproductive toxicity	Acute toxicity	Chronic (system) toxicity	Neurotoxicity	Environmental toxicity	Bioaccumulation	Persistence in the Environment
5-Nitro-o-anisidine	99-59-2	X									
Octachloronaphthalene	2234-13-1				X				X		
Osmium tetroxide	20816-12-0					X					
Parathion	56-38-2			X	X	X	X		X		
PCBs	1336-36-3	X									
Pentachloronitrobenzene (PCNB)	82-68-8	X		X	X				X	X	
Pentachlorophenol	87-86-5			X	X	X			X	X	
Peracetic acid	79-21-0					X	X				
Phenol	108-95-2			X		X			X	X	
p-Phenylenediamine	106-50-3		X			X					
2-Phenylphenol	90-43-7	X									
Phosgene	75-44-5					X					X
Phosphoric acid	7664-38-2										
Phosphorous (yellow or white)	7723-14-0				X	X	X		X		
Phthalic anhydride	85-44-9							X			
Picric acid	88-89-1								X		
1,3-Propane sultone	1120-71-4	X									
beta-Propiolactone	57-57-8	X	X			X					
Propionaldehyde	123-38-6						X		X		
Propxur	114-26-1						X				
Propylene	115-07-1										
Propylene dichloride	78-87-5	X							X		
Propylene oxide	75-56-9	X	X			X	X	X	X		X
Propyleneimine	75-55-8	X				X					
Pyridine	110-86-1								X		
Quinoline	91-22-5								X		
Saccharin	81-07-2	X	X	X	X						
Safrole	94-59-7	X			X		X				
Selenium	7782-49-2				X	X			X		
Sodium hydroxide (solution)	1310-73-2								X		
Sodium sulfate (solution)	7757-82-6								X		
Styrene	100-42-5	X	X				X		X		
Styrene oxide	96-09-3	X		X	X						
Sulfuric acid	7664-93-9					X	X		X		
Terephthalic acid	100-21-0						X	X			
1,1,2,2-Tetrachloroethane	79-34-5	X		X			X		X		
Tetrachloroethylene	127-18-4	X		X	X		X		X		X
Tetrachlorvinphos	961-11-5	X									
Thallium	7440-28-0										
Thioacetamide	62-55-5	X	X				X				
4,4'-Thiodianiline	139-65-1	X			X						
Thiourea	62-56-6	X		X	X	X					
Thorium dioxide	1314-20-1										
Titanium dioxide	13464-67-7										
Titanium tetrachloride	7550-45-0					X					
Toluene	108-88-3			X	X				X		
Toluenediamine	25376-45-8	X	X				X				
Toluene-2,4-diisocyanate	584-84-9					X					
Toluene-2,6-diisocyanate	91-08-7					X	X				
o-Toluidine hydrochloride	636-21-5	X									

Table C-1 Alphabetical Listing

Chemical Name	CAS Number	Carcinogenicity	Heritable genetic and chromosomal mutations	Developmental toxicity (including teratogenicity)	Reproductive toxicity	Acute toxicity	Chronic (system) toxicity	Neurotoxicity	Environmental toxicity	Bioaccumulation	Persistence in the Environment
o-Toluidine	95-53-4	X					X		X		
Toxaphene	8001-35-2	X		X	X	X	X	X	X		
Triaziquone	68-76-8	X	X		X						
Trichlorfon	52-68-6			X	X	X	X		X		X
1,2,4-Trichlorobenzene	120-82-1			X			X		X	X	
1,1,1-Trichloroethane	71-55-6			X	X				X		X
1,1,2-Trichloroethane	79-00-5	X					X		X		
Trichloroethylene	79-01-6	X		X	X		X				
2,4,5-Trichlorophenol	95-95-4						X		X		
2,4,6-Trichlorophenol	88-06-2	X							X		
Trifluralin	1582-09-8	X		X	X		X		X	X	
1,2,4-Trimethylbenzene	95-63-6								X		
Tris (2,3-dibromopropyl) phosphate	126-72-7	X	X	X	X						
Vanadium (fume or dust)	7440-62-2										
Vinyl acetate	108-05-4						X		X		
Vinyl bromide	593-60-2										
Vinyl chloride	75-01-4	X	X	X	X		X				
m-Xylene	108-38-3			X			X		X		
o-Xylene	95-47-6			X	X		X		X		
p-Xylene	106-42-3			X					X		
Xylene (mixed isomers)	1330-20-7			X	X		X		X		
2,6-Xylidine	87-62-7						X				
Zinc (fume or dust)	7440-66-6								X		
Zineb	12122-67-7								X		

Table C-2 Listing by CAS Number

Chemical Name	CAS Number	Carcinogenicity	Heritable genetic and chromosomal mutations	Developmental toxicity (including teratogenicity)	Reproductive toxicity	Acute toxicity	Chronic (system) toxicity	Neurotoxicity	Environmental toxicity	Bioaccumulation	Persistence in the Environment
Formaldehyde	50-00-0	X	X		X		X		X		
2,4-Dinitrophenol	51-28-5			X	X	X	X		X		
Nitrogen mustard	51-75-2	X	X	X	X	X					
Ethyl carbamate (urethane)	51-79-6	X	X								
Trichlorfon	52-68-6			X	X	X	X		X		X
2-Acetylaminofluorene	53-96-3	X									
N-Nitrosodiethylamine	55-18-5	X	X	X	X	X					
Benzamide	55-21-0										
Nitroglycerin	55-63-0								X		
Carbon tetrachloride	56-23-5	X		X			X		X		X
Parathion	56-38-2			X	X	X	X		X		
Cyanide compounds	57-12-5								X		
1,1-Dimethylhydrazine	57-14-7	X					X		X		
beta-Propiolactone	57-57-8	X	X				X				
Chlordane	57-74-9	X		X	X	X	X	X		X	
Lindane	58-89-9	X		X	X	X			X	X	
N-Nitrosomorpholine	59-89-2	X	X	X			X				

Table C-2 Listing by CAS Number

Chemical Name	CAS Number	Carcinogenicity	Heritable genetic and chromosomal mutations	Developmental toxicity (including teratogenicity)	Reproductive toxicity	Acute toxicity	Chronic (system) toxicity	Neurotoxicity	Environmental toxicity	Bioaccumulation	Persistence in the Environment
4-Aminoazobenzene	60-09-3	X									
4-Dimethylaminoazobenzene	60-11-7	X				X	X			X	
Methyl hydrazine	60-34-4	X				X	X		X		
Acetamide	60-35-5	X									X
Aniline	62-53-3	X				X	X		X		
Thioacetamide	62-55-5	X	X				X				
Thiourea	62-56-6	X		X	X	X					
Dichlorvis	62-73-7			X	X	X			X		X
N-Nitrosodimethylamine	62-75-9	X	X			X					
Carbaryl	63-25-2				X	X	X	X	X	X	X
Diethyl sulfate	64-67-5	X	X								
Methanol	67-56-1							X			
Isopropyl alcohol	67-63-0	X					X	X			
Acetone	67-64-1						X		X		
Chloroform	67-66-3	X		X	X	X			X		X
Hexachloroethane	67-72-1	X		X			X				
Triaziquone	68-76-8	X	X		X						
1-Butanol	71-36-3						X				
Benzene	71-43-2	X		X	X		X		X		
1,1,1-Trichloroethane	71-55-6			X	X				X		X
Methoxychlor	72-43-5			X	X			X	X	X	
Methyl bromide	74-83-9					X	X	X	X		X
Ethylene	74-85-1						X				
Methyl chloride	74-87-3			X	X		X				X
Methyl iodide	74-88-4	X						X			X
Hydrogen cyanide	74-90-8					X	X		X		
Methylene bromide	74-95-3							X			
Ethyl chloride	75-00-3										
Vinyl chloride	75-01-4	X	X	X	X		X				
Acetonitrile	75-05-8			X	X	X	X				X
Acetaldehyde	75-07-0	X				X			X		
Methylene chloride	75-09-2	X									X
Carbon disulfide	75-15-0			X	X		X	X	X		X
Ethylene oxide	75-21-8	X	X	X	X	X	X		X		X
Bromoform	75-25-2						X	X	X		
Dichlorobromomethane	75-27-4										X
1,1-Dichloroethylene	75-35-4	X		X	X		X		X		
Phosgene	75-44-5					X					X
Propyleneimine	75-55-8	X				X					
Propylene oxide	75-56-9	X	X			X	X	X	X		X
tert-Butyl alcohol	75-60-5										
Freon 113	76-13-1						X				
Heptachlor	76-44-8	X				X	X		X	X	
Hexachlorocyclopentadiene	77-47-4			X	X	X	X	X	X	X	X
Dimethyl sulfate	77-78-1	X	X			X			X		
Isobutyraldehyde	78-84-2						X				
Propylene dichloride	78-87-5	X							X		
sec-Butyl alcohol	78-92-2										
Methyl ethyl ketone (MEK)	78-93-3			X	X		X	X			
1,1,2-Trichloroethane	79-00-5	X					X		X		

Table C-2 Listing by CAS Number

Chemical Name	CAS Number	Carcinogenicity	Heritable genetic and chromosomal mutations	Developmental toxicity (including teratogenicity)	Reproductive toxicity	Acute toxicity	Chronic (system) toxicity	Neurotoxicity	Environmental toxicity	Bioaccumulation	Persistence in the Environment
Trichloroethylene	79-01-6	X		X	X		X				
Acrylamide	79-06-1		X		X	X	X	X	X		
Acrylic acid	79-10-7						X				
Chloroacetic acid	79-11-8					X					
Peracetic acid	79-21-0					X	X				
1,1,2,2-Tetrachloroethane	79-34-5	X		X			X		X		
Dimethylcarbomyl chloride	79-44-7	X				X					
2-Nitropropane	79-46-9	X		X	X	X	X				
4,4'-Isopropylidenediphenol	80-05-7										
Cumene hydroperoxide	80-15-9					X			X		
Methyl methacrylate	80-62-6				X		X				
Saccharin	81-07-2	X	X	X	X						
C.I. Food Red 15	81-88-9	X					X				
1-Amino-2-methylanthraquinone	82-28-0	X									
Pentachloronitrobenzene (PCNB)	82-68-8	X		X	X				X	X	
Diethylphthalate	84-66-2								X		
Dibutyl phthalate	84-74-2			X	X		X		X	X	
Phthalic anhydride	85-44-9							X			
Butyl benzyl phthalate	85-68-7						X		X		
N-Nitrosodiphenylamine	86-30-6	X							X		
2,6-Xylidine	87-62-7						X				
Hexachlorobutadiene	87-68-3	X		X	X	X	X		X	X	
Pentachlorophenol	87-86-5			X	X	X			X	X	
2,4,6-Trichlorophenol	88-06-2	X							X		
2-Nitrophenol	88-75-5						X	X	X		
Picric acid	88-89-1								X		
o-Anisidine	90-04-0	X					X	X			
2-Phenylphenol	90-43-7	X									
Michler's ketone	90-94-8	X									
Toluene-2,6-diisocyanate	91-08-7					X	X				
Naphthalene	91-20-3								X	X	
Quinoline	91-22-5								X		
beta-Naphthylamine	91-59-8	X									
3,3'-Dichlorobenzidine	91-94-1	X									
Biphenyl	92-52-4			X			X		X		
4-Aminobiphenyl	92-67-1	X									
Benzidine	92-87-5	X				X	X				
4-Nitrobiphenyl	92-93-3	X									
Benzoyl peroxide	94-36-0										
Safrole	94-59-7	X			X		X				
2,4-D	94-75-7			X	X	X	X		X		
o-Xylene	95-47-6			X	X		X		X		
o-Cresol	95-48-7					X			X		
o-Dichlorobenzene	95-50-1								X	X	
o-Toluidine	95-53-4	X					X		X		
1,2,4-Trimethylbenzene	95-63-6								X		
2,4-Diaminotoluene	95-80-7	X	X								
2,4,5-Trichlorophenol	95-95-4						X		X		
Styrene oxide	96-09-3	X		X	X						
1,2-Dibromo-3-chloropropane	96-12-8	X		X	X	X	X		X		

Chemical Name	CAS Number	Carcinogenicity	Heritable genetic and chromosomal mutations	Developmental toxicity (including teratogenicity)	Reproductive toxicity	Acute toxicity	Chronic (system) toxicity	Neurotoxicity	Environmental toxicity	Bioaccumulation	Persistence in the Environment
Methyl acrylate	96-33-3								X		
Ethylenethiourea	96-45-7	X	X	X	X		X				
C.I. Solvent Yellow 3	97-56-3	X								X	
Benzotrichloride	98-07-7	X					X				
1-Methyl ethyl benzene (Cumene)	98-82-8						X	X	X		
Benzal chloride	98-87-3	X				X					
Benzoyl chloride	98-88-4								X		
Nitrobenzene	98-95-3				X	X	X		X		
5-Nitro-o-anisidine	99-59-2	X									
p-Nitrophenol	100-02-7					X	X	X	X		
Terephthalic acid	100-21-0						X	X			
Ethylbenzene	100-41-4			X	X		X		X		
Styrene	100-42-5	X	X				X		X		
Benzyl chloride	100-44-7	X		X		X	X	X	X		
N-Nitrosopiperidine	100-75-4	X	X			X	X				
4,4'-Methylenebis (2-chloroaniline)	101-14-4	X									
4,4'-Methylenebis (N,N-dimethyl)	101-61-1	X									
MBI	101-68-8										
4,4'-Methylenedianiline	101-77-9						X				
4,4'-Diaminodiphenyl ether	101-80-4	X					X				
Bis (2-ethylhexyl) adipate	103-23-1	X	X								
p-Anisidine	104-94-9										
2,4-Dimethylphenol	105-67-9								X		
p-Xylene	106-42-3			X					X		
p-Cresol	106-44-5					X			X		
p-Dichlorobenzene	106-46-7	X					X		X	X	
p-Phenylenediamine	106-50-3		X			X					
p-Benzoquinone	106-51-4					X			X		
1,2-Butylene oxide	106-89-7										
Epichlorohydrin	106-89-8	X	X		X	X	X		X		X
Ethylene dibromide	106-93-4	X	X	X	X	X	X		X		
1,3-Butadiene	106-99-0	X		X	X		X				
Acrolein	107-02-8					X	X		X		
Allyl chloride	107-05-1	X		X		X	X	X	X		
Ethylene dichloride	107-06-2	X	X								X
Acrylonitrile	107-13-1	X		X	X	X			X		
Ethylene glycol	107-21-1						X				
Chloromethyl methyl ether	107-30-2	X				X					
Vinyl acetate	108-05-4						X		X		
Methyl isobutyl ketone	108-10-1						X	X			
Maleic anhydride	108-31-6										
m-Xylene	108-38-3			X			X		X		
m-Cresol	108-39-4					X			X		
Dichloroisopropyl ether	108-60-1	X					X	X			
Melamine	108-78-1	X			X						
Toluene	108-88-3			X	X				X		
Chlorobenzene	108-90-7						X		X		
Phenol	108-95-2			X		X			X	X	
2-Methoxyethanol	109-86-4			X	X		X				
Ethylene glycol monoethyl ether	110-80-5			X	X		X				

Table C-2 Listing by CAS Number

Chemical Name	CAS Number	Carcinogenicity	Heritable genetic and chromosomal mutations	Developmental toxicity (including teratogenicity)	Reproductive toxicity	Acute toxicity	Chronic (system) toxicity	Neurotoxicity	Environmental toxicity	Bioaccumulation	Persistence in the Environment
Cyclohexane	110-82-7								X		
Pyridine	110-86-1								X		
Diethanolamine	111-42-2	X							X		
Dichloroethyl ether	111-44-4	X				X	X				
Propxur	114-26-1						X				
Propylene	115-07-1										
Dicofol	115-32-2	X				X			X		
2-Aminoanthraquinone	117-79-3	X					X				
Diethylhexyl phthalate	117-81-7	X	X	X	X		X		X	X	
Di-n-octylphthalate	117-84-0					X	X			X	
Hexachlorobenzene	118-74-1	X		X	X	X	X		X	X	
3,3'-Dimethoxybenzidine	119-90-4	X									
3,3'-Dimethylbenzidine	119-93-7	X									
Anthracene	120-12-7						X		X	X	
p-Cresidine	120-71-8	X									
Catechol	120-80-9					X			X		
1,2,4-Trichlorobenzene	120-82-1			X			X		X	X	
2,4-Dichlorophenol	120-83-2								X		
2,4-Dinitrotoluene	121-14-2	X					X		X		
N,N-Dimethylaniline	121-69-7						X	X			
1,2-Diphenylhydrazine	122-66-7	X							X		
Hydroquinone	123-31-9					X	X		X		
Propionaldehyde	123-38-6						X		X		
Butyraldehyde	123-72-8								X		
Dioxane	123-91-1	X									
Tris (2,3-dibromopropyl) phosphate	126-72-7	X	X	X	X						
Chloroprene	126-99-8		X	X	X		X	X			
Tetrachloroethylene	127-18-4	X		X	X		X		X		X
C.I. Vat Yellow 4	128-66-5	X									
Dimethylphthalate	131-11-3					X			X		
Dibenzofuran	132-64-9										
Captan	133-06-2	X	X	X	X	X			X		X
Chloramben	133-90-4	X									
o-Anisidine hydrochloride	134-29-2	X									
alpha-Naphthylamine	134-32-7	X									
Cupferron	135-20-6	X									
Nitrilotriacetic acid	139-13-9	X					X				
4,4'-Thiodianiline	139-65-1	X			X						
Ethyl acrylate	140-88-5	X		X			X		X		X
Butyl acrylate	141-32-2								X		
Ethyleneimine	151-56-4	X	X		X	X	X		X		X
p-Nitrosodiphenylamine	156-10-5	X									
Calcium cyanamide	156-62-7					X	X				
Hydrazine	302-01-2	X				X			X		
Aldrin	309-00-2	X		X	X	X	X		X	X	
Diazomethane	334-88-3										
Carbonyl sulfide	463-58-1						X				
Auramine	492-80-8	X									
Mustard gas	505-60-2	X	X			X					
Chlorobenzilate	510-15-6	X							X		

Table C-2 Listing by CAS Number

Chemical Name	CAS Number	Carcinogenicity	Heritable genetic and chromosomal mutations	Developmental toxicity (including teratogenicity)	Reproductive toxicity	Acute toxicity	Chronic (system) toxicity	Neurotoxicity	Environmental toxicity	Bioaccumulation	Persistence in the Environment
2-Chloroacetophenone	532-27-4					X					
4,6-Dinitro-o-cresol	534-52-1					X			X		
Ethyl chloroformate	541-41-3					X					
m-Dichlorobenzene	541-73-1								X		
1,3-Dichloropropene	542-75-6	X				X	X		X		
Dichloromethyl ether	542-88-1	X				X					
C.I. Basic Green 4	569-64-2					X			X		
Toluene-2,4-diisocyanate	584-84-9					X					
Vinyl bromide	593-60-2										
2,6-Dinitrotoluene	606-20-2					X			X		
2,4-Diaminoanisole	615-05-4	X									
Di-n-propylnitrosamine	621-64-7	X									
Methyl isocyanate	624-83-9					X					
o-Toluidine hydrochloride	636-21-5	X									
Hexamethylphosphoramide	680-31-9	X	X				X				
N-Nitroso-N-methylurea	684-93-5	X	X	X		X					
N-Nitroso-N-ethyl urea	759-73-9	X	X	X	X						
C.I. Solvent Yellow 14	842-07-9	X									
N-Nitrosodi-n-butylamine	924-16-3	X									
Tetrachlorvinphos	961-11-5	X									
C.I. Basic Red 1	989-38-8	X									
1,3-Propane sultone	1120-71-4	X									
Decabromodiphenyl oxide	1163-19-5	X									
Maneb	1247-38-2			X	X		X	X	X		
Sodium hydroxide (solution)	1310-73-2								X		
Molybdenum trioxide	1313-27-5					X	X	X			
Thorium dioxide	1314-20-1										
Cresol (mixed isomers)	1319-77-3					X	X		X		
Xylene (mixed isomers)	1330-20-7			X	X		X		X		
Asbestos (friable)	1332-21-4	X					X				
Hexachloronaphthalene	1335-87-1										
PCBs	1336-36-3	X									
Aluminum oxide	1344-28-1						X				
1:2,3:4-Diepoxybutane	1464-53-5	X	X			X					
Trifluralin	1582-09-8	X		X	X		X		X	X	
Methyl tert-butyl ether	1634-04-4										
Nitrofen	1836-75-5	X		X	X		X	X	X		
Chlorothalonil	1897-45-6	X							X		
C.I. Direct Black 38	1937-37-7	X									
Fluometuron	2164-17-2	X					X		X		
Octachloronaphthalene	2234-13-1				X				X		
Diallate	2303-16-4	X			X				X		
C.I. Direct Blue 6	2602-46-2	X		X							
C.I. Acid Blue 9, diammonium salt	2650-18-2										
C.I. Disperse Yellow 3	2832-40-8	X									
C.I. Solvent Orange 7	3118-97-6										
C.I. Food Red 5	3761-53-3	X									
C.I. Acid Blue 9, disodium salt	3844-45-9	X									
N-Nitrosomethylvinylamine	4549-40-0	X	X			X					
C.I. Acid Green 3	4680-78-8	X									

Table C-2 Listing by CAS Number

Chemical Name	CAS Number	Carcinogenicity	Heritable genetic and chromosomal mutations	Developmental toxicity (including teratogenicity)	Reproductive toxicity	Acute toxicity	Chronic (system) toxicity	Neurotoxicity	Environmental toxicity	Bioaccumulation	Persistence in the Environment
Ammonium nitrate (solution)	6484-52-2						X				
Aluminum (fume or dust)	7429-90-5										
Lead	7439-92-1			X	X			X			
Manganese	7439-96-5							X			
Mercury	7439-97-6				X		X	X	X		
Nickel	7440-02-0	X		X	X		X		X		
Thallium	7440-28-0										
Antimony	7440-36-0				X						
Arsenic	7440-38-2	X									
Barium	7440-39-3										
Beryllium	7440-41-7	X					X				
Cadmium	7440-43-9	X		X	X	X	X		X		
Chromium	7440-47-3	X					X		X		
Cobalt	7440-48-4						X				
Copper	7440-50-8			X	X				X		
Vanadium (fume or dust)	7440-62-2										
Zinc (fume or dust)	7440-66-6								X		
Titanium tetrachloride	7550-45-0					X					
Hydrochloric acid	7647-01-0					X	X				
Phosphoric acid	7664-38-2										
Hydrogen fluoride	7664-39-3		X	X	X	X	X				
Ammonia	7664-41-7					X	X			X	
Sulfuric acid	7664-93-9					X	X		X		
Nitric acid	7697-37-2					X					
Phosphorous (yellow or white)	7723-14-0				X	X	X		X		
Sodium sulfate (solution)	7757-82-6								X		
Selenium	7782-49-2			X	X				X		
Chlorine	7792-50-5					X	X		X		
Ammonium sulfate (solution)	7783-20-2								X		
Toxaphene	8001-35-2	X		X	X	X	X	X	X		
Hydrazine sulfate	10034-93-2	X									
Chlorine dioxide	10049-04-4			X	X						
Zineb	12122-67-7								X		
Titanium dioxide	13464-67-7										
C.I. Direct Brown 95	16071-86-6	X									
N-Nitrosonornicotine	16543-55-8	X									
Osmium tetroxide	20816-12-0						X				
Dichlorobenzene	25321-22-6	X						X	X	X	
Toluenediamine	25376-45-8	X	X					X			
2,4-Diaminoanisole sulfate	39156-41-7	X									

Source for Tables C-1 and C-2: Information from the United States Environmental Protection Agency, published by Greenpeace in *A Citizen's Toxic Waste Audit Manual* (see Resources for Chapter 5).

Appendix D

Other Useful Right to Know Information

In addition to the Toxics Release Inventory, the 1986 Superfund Amendments and Reauthorization Act requires industry to provide information to State Emergency Response Commissions (SERCs) and Local Emergency Planning Committees (LEPCs). Although neither of these bodies collects data on the quantities of routine emissions from plants, they do collect information that might be a useful supplement to the TRI data discussed in Appendix A.

Chemical inventories. State Emergency Response Commissions, Local Emergency Planning Committees, and local fire departments maintain Material Safety Data Sheets (MSDS) for facilities required to complete them under the federal Occupational Safety and Health Act's worker right to know standard. These data sheets are designed to inform the workers in a plant about the health risks of workplace exposure to specific chemicals used in the plant. About 50,000 chemicals are included in this requirement.

As noted in Chapter 5, community exposure differs from worker exposure because communities may be exposed 24 hours a day, while workers are usually exposed only 8 hours a day. In addition, communities may contain particularly sensitive populations, such as the elderly and the very young. For these reasons, the MSDS information cannot be used directly to understand community risks, but it may raise concerns about particular chemicals that you can explore with company representatives.

Emergency planning information. The federal Extremely Hazardous Substance List contains over 360 chemicals that could cause serious human health effects from short-term exposures such as accidental air releases. A Local Emergency Planning Committee can request material necessary to do emergency planning (see forms) for any facility using or storing these extremely hazardous substances above threshold amounts. This material includes safety audits, internal hazard assessments, insurance material, source reduction information, and more. Keep this information in mind not only as a source for collecting background data, but also as a mechanism for getting information from an uncooperative company, if necessary.

Resources

OMB Watch. *Using Community Right to Know: A Guide to a New Federal Law.* (Available from OMB Watch, 2001 O Street NW, Washington DC 20036. (202) 659-1711.)

This document contains a complete description of the provisions outlined in this chapter.

Tier Two

EMERGENCY AND HAZARDOUS CHEMICAL INVENTORY

Specific Information by Chemical

Facility Identification

Name _____
Street Address _____
City _____ , State ____ Zip ____

SIC Code _____   Dun & Brad Number ____ - ____

FOR OFFICIAL USE ONLY
ID # ____
Date Received ____

Owner/Operator Name

Name _____
Mail Address _____
Phone (____) ____

Emergency Contact

Name _____
Title _____
Phone (____)   24 Hr. Phone (____)

Name _____
Title _____
Phone (____)   24 Hr. Phone (____)

Important: Read all instructions before completing form

Reporting Period From January 1 to December 31, 19___

Chemical Description	Physical and Health Hazards (check all that apply)	Inventory				Storage Codes and Locations (Non-Confidential)	
		Max. Daily Amount (code)	Avg. Daily Amount (code)	No. of Days On-site (days)		Storage Code	Storage Locations

Chemical Description

CAS _____   Trade Secret ☐
Chem. Name _____
Check all that apply: ☐ Pure ☐ Mix   ☐ Solid ☐ Liquid ☐ Gas

Physical and Health Hazards (check all that apply):
☐ Fire
☐ Sudden Release of Pressure
☐ Reactivity
☐ Immediate (acute)
☐ Delayed (chronic)

CAS _____   Trade Secret ☐
Chem. Name _____
Check all that apply: ☐ Pure ☐ Mix   ☐ Solid ☐ Liquid ☐ Gas

Physical and Health Hazards (check all that apply):
☐ Fire
☐ Sudden Release of Pressure
☐ Reactivity
☐ Immediate (acute)
☐ Delayed (chronic)

CAS _____   Trade Secret ☐
Chem. Name _____
Check all that apply: ☐ Pure ☐ Mix   ☐ Solid ☐ Liquid ☐ Gas

Physical and Health Hazards (check all that apply):
☐ Fire
☐ Sudden Release of Pressure
☐ Reactivity
☐ Immediate (acute)
☐ Delayed (chronic)

Certification (Read and sign after completing all sections)

I certify under penalty of law that I have personally examined and am familiar with the information submitted in this and all attached documents, and that based on my inquiry of those individuals responsible for obtaining the information, I believe that the submitted information is true, accurate, and complete.

Name and official title of owner/operator OR owner/operator's authorized representative _____

Signature _____   Date signed _____

Optional Attachments (Check one)
☐ I have attached a site plan
☐ I have attached a list of site coordinate abbreviations

Tier Two

EMERGENCY AND HAZARDOUS CHEMICAL INVENTORY

Specific Information by Chemical

Facility Identification

Name _____

Street Address _____

City _____ State _____ Zip _____

SIC Code [ ][ ][ ][ ]   Dun & Brad Number [ ][ ][ ][ ]-[ ][ ][ ][ ][ ]

FOR OFFICIAL USE ONLY

ID # [ ][ ][ ]

Date Received [ ]

Owner/Operator Name

Name _____

Mail Address _____

Phone ( ) _____

Emergency Contact

Name _____   Title _____

Phone ( ) _____   24 Hr. Phone ( ) _____

Name _____   Title _____

Phone ( ) _____   24 Hr. Phone ( ) _____

Reporting Period From January 1 to December 31, 19 ___

Important: Read all instructions before completing form

Confidential Location Information Sheet

Storage Codes and Locations
(Confidential)

Storage Codes	*Storage Locations*

CAS # [ ][ ][ ][ ][ ][ ][ ][ ] [ ] [ ] Chem. Name

[ ][ ][ ][ ][ ]	_____
[ ][ ][ ][ ][ ]	_____
[ ][ ][ ][ ][ ]	_____

CAS # [ ][ ][ ][ ][ ][ ][ ][ ] [ ] [ ] Chem. Name

[ ][ ][ ][ ][ ]	_____
[ ][ ][ ][ ][ ]	_____
[ ][ ][ ][ ][ ]	_____

CAS # [ ][ ][ ][ ][ ][ ][ ][ ] [ ] [ ] Chem. Name

[ ][ ][ ][ ][ ]	_____
[ ][ ][ ][ ][ ]	_____
[ ][ ][ ][ ][ ]	_____

Certification *(Read and sign after completing all sections)*

I certify under penalty of law that I have personally examined and am familiar with the information submitted in this and all attached documents, and that based on my inquiry of those individuals responsible for obtaining the information, I believe that the submitted information is true, accurate, and complete.

_____   _____   _____
Name and official title of owner/operator OR owner/operator's authorized representative   Signature   Date signed

Optional Attachments *(Check one)*

[ ] I have attached a site plan

[ ] I have attached a list of site coordinate abbreviations

Federal Waste Minimization Forms

The United States Environmental Protection Agency's Office of Solid Waste includes a hazardous waste minimization program. It focuses entirely on strategies to reduce the volume of Resource Conservation and Recovery Act (RCRA) hazardous wastes shipped from plant sites. These strategies include recycling as well as source reduction.

Minimization programs have not focused on reducing air and water pollution produced by plants — wastes that INFORM's research has shown may be at least as great in quantity as those regulated under RCRA. For example, INFORM analyses of discharges of eight organic chemical plants in New Jersey revealed that 35 percent of chemical wastes were generated as solid wastes, 23 percent as air pollutants, and 42 percent as water pollutants.

Minimization programs are not only limited by their focus on RCRA wastes. They also tend to make little use of the most environmentally beneficial management strategy — waste reduction at source. Instead, they rely mainly on strategies that involve management of wastes already created, strategies entailing generally more expensive engineering-oriented pollution control solutions that corporate engineers and government regulators have traditionally used.

Facilities that generate more than 220 pounds per month of RCRA hazardous wastes are required to submit waste minimization reports in odd-numbered years. While the waste minimization program has limitations, the forms that companies fill out can provide background information on a plant's source reduction program that is useful to review before a plant visit. The forms are available from state or federal EPA offices, depending on whether the state has the authority to implement RCRA regulations. Copies of the 1987 form are shown here; the 1989 forms have been organized somewhat differently.

U.S. ENVIRONMENTAL PROTECTION AGENCY

1987 Hazardous Waste Report

WASTE MINIMIZATION

FORM WM

PART I

BEFORE COPYING FORM, ATTACH SITE IDENTIFICATION LABEL
OR ENTER:

SITE NAME _____

EPA ID NO. _____

WHO MUST COMPLETE THIS FORM?

Form WM Part I, describing efforts undertaken to implement waste minimization programs, must be completed by all generators required to file an Annual/Biennial Report. This requirement was established in response to statutory provisions included in the Hazardous and Solid Waste Amendments of 1984 (HSWA).

NOTE: Generators shipping hazardous waste off site are required to certify, on Item 16 of the Uniform Hazardous Waste Manifest, that they have a program in place to reduce, to the degree determined economically practicable, the volume and toxicity of hazardous waste generated. A similar certification must also be made by generators who have obtained a RCRA treatment, storage, or disposal permit. Consistent with these certification requirements, generators must report, on Form WM Part I, the efforts undertaken to implement waste minimization programs.

INSTRUCTIONS: Please read the detailed instructions on page 8 of the 1987 Hazardous Waste Report Instruction booklet before completing this form.

Answer questions 1 through 10. Throughout this form enter "DK" if the information requested is not known or is not available; enter "NA" if the information requested is not applicable.

1. Did this site create or expand a source reduction and recycling program?

	1987		1986		Prior Years	
	Yes	No	Yes	No	Yes	No
Create	☐	☐	☐	☐	☐	☐
Expand	☐	☐	☐	☐	☐	☐

2. Did this site have a written policy or statement that outlined goals, objectives and methods for source reduction and recycling of hazardous waste?

	1987		1986		Prior Years	
	Yes	No	Yes	No	Yes	No
Yes	☐		☐		☐	
No	☐		☐		☐	

3. What was the dollar amount of capital expenditures (plant and equipment) and operating costs devoted to source reduction and recycling of hazardous waste? ENTER ZERO (0) IF NONE.

	1987	1986	Prior Years
Capital expenditures	$ _____	$ _____	$ _____
Operating costs	$ _____	$ _____	$ _____

4. Did this site have an employee training program or provide incentives (bonuses, awards, personal recognition, etc.) to identify and implement source reduction and recycling opportunities and activities?

	1987		1986		Prior Years	
	Yes	No	Yes	No	Yes	No
Training	☐	☐	☐	☐	☐	☐
Incentives	☐	☐	☐	☐	☐	☐

5. Did this site conduct a source reduction and/or recycling opportunity assessment or audit? Note: an opportunity assessment or audit is a procedure that identifies practices that can be implemented to reduce the generation of hazardous waste or the quantity which must subsequently be treated, stored or disposed.

	1987		1986		Prior Years	
	Yes	No	Yes	No	Yes	No
Site-Wide	☐	☐	☐	☐	☐	☐
Process-Specific	☐	☐	☐	☐	☐	☐

6. Did this site identify or implement new SOURCE REDUCTION opportunities to reduce the volume and/or toxicity of hazardous waste generated at this site?

	1987		1986		Prior Years	
	Yes	No	Yes	No	Yes	No
Identify	☐	☐	☐	☐	☐	☐
Implement	☐	☐	☐	☐	☐	☐

7. What factors have delayed or prevented implementation of SOURCE REDUCTION opportunities. MARK ☒ NEXT TO ALL THAT APPLY.

- a. ☐ Insufficient capital to install new source reduction equipment or implement new source reduction practices.
- b. ☐ Lack of technical information on source reduction techniques, applicable to my specific production processes.
- c. ☐ Source reduction is not economically feasible: cost savings in waste management or production will not recover the capital investment.
- d. ☐ Concern that product quality may decline as a result of source reduction.
- e. ☐ Technical limitations of the production processes.
- f. ☐ Permitting burdens.
- g. ☐ Other (SPECIFY) _____

8. Did this site identify or implement new RECYCLING opportunities to reduce the volume and/or toxicity of hazardous waste generated at this site or subsequently treated, stored, or disposed of on site or off site?

	1987		1986		Prior Years	
	Yes	No	Yes	No	Yes	No
Identify	☐	☐	☐	☐	☐	☐
Implement	☐	☐	☐	☐	☐	☐

FORM WM - PART I

9. What factors have delayed or prevented implementation of on-site or off-site RECYCLING opportunities. MARK ☒ NEXT TO ALL THAT APPLY.

☐ a. Insufficient capital to install new recycling equipment or implement new recycling practices.

☐ b. Lack of technical information on recycling techniques applicable to this site's specific production processes.

☐ c. Recycling is not economically feasible: cost savings in waste management or production will not recover the capital investment.

☐ d. Concern that product quality may decline as a result of recycling.

☐ e. Requirements to manifest wastes inhibit shipments off site for recycling.

☐ f. Financial liability provisions inhibit shipments off site for recycling.

☐ g. Technical limitations of product processes inhibit shipments off site for recycling.

☐ h. Technical limitations of production processes inhibit on-site recycling.

☐ i. Permitting burdens inhibit recycling.

☐ j. Lack of permitted off-site recycling facilities.

☐ k. Unable to identify a market for recyclable materials.

☐ l. Other (SPECIFY) _____

10. Has this site requested or received technical information or financial assistance on source reduction and/or recycling practices from any of the following sources? MARK ☒ NEXT TO ALL THAT APPLY.

	1987 Technical	1987 Financial	1986 Technical	1986 Financial	Prior Years Technical	Prior Years Financial
a. Local government	☐	☐	☐	☐	☐	☐
b. State government	☐	☐	☐	☐	☐	☐
c. Federal government	☐	☐	☐	☐	☐	☐
d. Trade associations	☐	☐	☐	☐	☐	☐
e. Educational institutions	☐	☐	☐	☐	☐	☐
f. Suppliers	☐	☐	☐	☐	☐	☐
g. Other parts of your firm	☐	☐	☐	☐	☐	☐
h. Other firms/consultants	☐	☐	☐	☐	☐	☐
i. No request made	☐	☐	☐	☐	☐	☐
j. Other (conferences, literature, etc.)	☐	☐	☐	☐	☐	☐

Comments:

BEFORE COPYING FORM, ATTACH SITE IDENTIFICATION LABEL OR ENTER:

SITE NAME _____

EPA ID NO. _____

WHO MUST COMPLETE THIS FORM?

Form WM Part II must be completed only by generators that engaged in an activity during 1987 that resulted in waste minimization.

Waste minimization means:
(1) reduction in the volume and/or toxicity of hazardous waste generated as a result of source reduction; and/or,
(2) reduction in the volume and/or toxicity of hazardous waste subsequently treated, stored, or disposed as a result of on-site or off-site recycling.

Mark ☒ and do not complete this form if no waste minimization results were achieved during 1987. ☐

INSTRUCTIONS:

Please read the detailed instructions on page 10 of the 1987 Hazardous Waste Report Instruction booklet before completing this form.

Make and complete a photocopy of this form for each hazardous waste minimized in 1987.

Complete Sections I through IV. Throughout this form enter "DK" if the information requested is not known or is not available; enter "NA" if the information requested is not applicable.

Sec. I

A. EPA hazardous waste code — Instruction Page 11

B. State hazardous waste code — Page 11

C. Product or service description — Page 11

D. Product or service SIC code — Page 11

E. Waste form code — Page 11

F. UOM — Page 12

G. Density — Page 12 ☐☐.☐ lbs/gal ☐ kg

H. Source description: — Page 12

I. Source code — Page 12

Sec. II

A. 1986 quantity generated — Instruction Page 13

B. 1987 quantity generated — Page 13

C. Production ratio — Page 13

D. Toxicity change code — Page 15

E. Waste minimization: recycling — Page 15
Code
1. ☐
2. ☐

Quantity recycled

F. Waste minimization: source reduction — Page 16
Code
1. ☐
2. ☐
3. ☐

Quantity prevented

Sec. III

A. Narrative description of waste minimization project or activity and results achieved — Instruction Page 23

FORM WM - PART II

Sec. IV.	**Instructions:** Answer questions 1 through 4. Mark ☒ next to **the effects produced by the source reduction and/or recycling activity** reported on this form in Sections I through III.

1. What effect did this site's source reduction and/or recycling activity have on the **quantity of water effluent** produced by hazardous waste generation processes during 1987?

   ☐ a. Increase in the quantity of water effluent
   ☐ b. Decrease in the quantity of water effluent
   ☐ c. No effect on the quantity of water effluent
   ☐ d. Don't know

2. What effect did this site's source reduction and/or recycling activity have on the **toxicity of water effluent** produced by hazardous waste generation processes during 1987?

   ☐ a. Increase in the concentration of hazardous constituents
   ☐ b. Decrease in the concentration of hazardous constituents
   ☐ c. No effect on the concentration of hazardous constituents
   ☐ d. Don't know

3. What effect did this site's source reduction and/or recycling activity have on the **quantity of air emissions** produced by hazardous waste generation processes during 1987?

   ☐ a. Increase in the quantity of air emissions
   ☐ b. Decrease in the quantity of air emissions
   ☐ c. No effect on the quantity of air emissions
   ☐ d. Don't know

4. What effect did this site's source reduction and/or recycling activity have on the **toxicity of the air emissions** produced by hazardous waste generation processes during 1987?

   ☐ a. Increase in the concentration of hazardous constituents
   ☐ b. Decrease in the concentration of hazardous constituents
   ☐ c. No effect on the concentration of hazardous constituents
   ☐ d. Don't know

Comments:

Appendix F

Glossary

Batch processing. Production of different products at different times, using the same equipment, to meet changing customer needs. (See **continuous processing**.)

CAS number. The American Chemical Society's Chemical Abstract Services unique number for each chemical; a given chemical may have more than one name but only one CAS number.

Chemical-specific. Applying to individual chemicals, versus broad classes of chemicals (such as chemical-specific data, versus waste stream data, referring to a combination of chemicals as one unit).

Continuous processing. Production in which process equipment is dedicated to continuous production of a single product. Continuous process operations usually generate less waste per pound of product than do batch process operations. (See **batch processing**.)

Deep well injection. Disposal of hazardous waste deep underground, presumably below groundwater supplies; considered by many to be a very dangerous and undesirable method of waste management. (Also called underground injection.)

End-of-the-pipe. Applying to regulations or pollution control strategies that take place after all products and waste products have been made and the waste products are released from the plant.

Facility-level materials accounting. A comparison of the inputs of individual chemicals (the amount existing in plant inventory plus the amount entering the plant) with the outputs (the amount consumed in plant processes plus the amount shipped out in products) to determine the amounts released to the environment.

Form R. Form on which companies report Toxics Release Inventory data to state and federal environmental officials (see Appendix A).

Fugitive air emissions. Air pollutants released through leaky valves, evaporation from tanks, and other *unintentional* release points.

Hazardous waste. According to federal environmental law *specifically* refers to solid hazardous discharges (not air pollutants) that are regulated under the Resource Conservation and Recovery Act.

Materials balance. Quantitative assessment of chemical inputs and outputs of individual processes that aims to account for every pound of a chemical (a) shipped to the process, (b) created or destroyed in the process, (c) delivered as a product from the process, and (d) released as waste; useful as a tool for determining if all sources of waste have been identified.

Multimedia. Applying to all environmental media: land, water, and air.

POTW. See Publicly owned treatment works.

Point source air emissions. Air pollutants released through smokestacks, vents, and other *intentional* release points.

Pollution prevention. As used by the federal Environmental Protection Agency, source reduction *or* environmentally sound recycling.

Process-level source reduction inventory. A series of steps enabling internal company management to identify the sources and activities, on an individual process basis, that lead to releases of waste to the air, water, and land.

Publicly owned treatment works (POTW). Public sewage facilities.

RCRA. See Resource Conservation and Recovery Act.

Recycling. The reuse of by-products, or reclamation of components of by-products, that might otherwise be disposed of in the environment. As currently used by the public at large, recycling includes everything from the creation of new paper from old paper to the burning of used oil and chemicals in furnaces. Some find it helpful to distinguish environmentally sound recycling from other recycling practices that may not be desirable.

Resource Conservation and Recovery Act (RCRA). The federal "cradle to grave" regulations for solid waste, both hazardous and nonhazardous (garbage).

SARA. See Superfund Amendments and Reauthorization Act.

SIC codes. Standard Industrial Classification codes by which the federal government classifies United States companies by their products.

Source reduction. According to the congressional Office of Technology Assessment, "In-plant practices that reduce, avoid or eliminate the generation of hazardous waste so as to reduce risks to health and environment." This includes recycling that is "an integral part of the . . . industrial process," or "in-process" recycling, but not recycling that involves moving wastes to another area of the plant, or to an off-site location.

Superfund. The federal law that regulates and raises money for the clean-up of hazardous waste dumps such as Love Canal, Stringfellow, and Times Beach.

Superfund Amendments and Reauthorization Act (SARA). Amendments to the original Superfund law, including the right-to-know provisions such as the Toxics Release Inventory requirements.

Toxic wastes. Specific chemicals or chemical mixtures that are thought to be injurious to human or environmental health, regardless of regulatory status or environmental medium into which they are released.

Waste minimization. As defined in a 1986 Environmental Protection Agency report to Congress, "the reduction, to the extent feasible, of hazardous waste that is generated or subsequently treated, stored, or disposed of. It includes any source reduction or recycling activity undertaken by a generator that results in either (1) the reduction of total volume or quantity of hazardous waste, or (2) the reduction of toxicity of hazardous waste, or both, so long as such reduction is consistent with the goal of minimizing present and future threats to human health and the environment." It is important to realize that waste minimization is not a multimedia approach. It is concerned only with hazardous wastes as defined under the Resource Conservation and Recovery Act; this includes solid hazardous wastes and certain wastewaters destined for land disposal, but does not include air emissions.

Appendix G

Worksheets

Plant Background Information

Plant: _____

Date Filled Out: _____

A. Identification

i. Plant name _____

ii. Plant ownership or affiliation (include most recent changes in ownership) _____

_____

B. Output and employment

i. Annual sales data (most recent year) _____

ii. Annual production volume (most recent year)* _____

iii. Number of plant employees in (a) administration _____

and (b) production _____

iv. Date operations began_____

v. Major changes in output and type of product in the past ten years _____

_____

_____

C. Product type

i. Categories of products manufactured at the plant (include Standard Industrial Code (SIC) for product type; these codes are listed in Table A-1 in Appendix A)

_____

_____

_____

ii. Type of production process used: continuous or batch (**Continuous** processes are dedicated to continuous production of a single product; **batch** processes produce different products at different times. Continuous operations usually generate less waste per pound of product.)*

_____

iii. Changes in production output planned that might proportionally affect the quantity of toxic releases*

_____

_____

*May not be available in public databases; this information may have to be obtained during plant visit.

Plant Background Information	**Worksheet 2**
Plant: _____ Date Filled Out: _____	*Permit Compliance and Applications*

A. Environmental Non-Compliance Profile

Wastewater Discharge (specify pollutants)

Pollutant	Description of Violation

Air Emission (specify pollutants)

Pollutant	Description of Violation

Hazardous Waste Treatment, Storage, Disposal Facility (TSD) Permit *

Facility Type/ Capacity	Description of Violation	Operating/ Not Operating

*Also required for hazardous waste recycling facilities

B. Violations

i. Circumstances in which the plant has exceeded limits established in its permit.

ii. If the violation was sudden or accidental, what steps has the plant taken to prevent a recurrence?

iii. Have any permit violations led the plant to take a closer look at source reduction?

C. Permit Applications

i. Does the plant have any applications pending before state or local government for modifications to any existing permits? If so, what are they?

ii. Does the plant have any applications pending before the state or local government for a new permit (e.g., construction of a new incinerator or landfill)? If so, what are they?

iii. In reviewing the plant's application for a permit modification or new permit, has the state inquired whether a source reduction program exists, or required one? If so, explain.

iv. Has the plant submitted a premanufacturer's notification (PMN) for any new products or processes as required by the Toxic Substance Control Act (TSCA)? (Requests for TSCA related information should be sent to EPA headquarters, ATTN: Jeralene Green (A-101).)

Worksheet 3
Plant Pollution Profile

(All amounts in pounds per year)

Plant Background Information

Plant: _____

Date Filled Out: _____

Pollutant Name/CAS Number				
Total Releases	Before end-of-pipe management			
	After end-of-pipe management			
Air	Fugitive emissions			
	Stack emissions			
	Total Released to Air			
Water	Released to surface water			
	Underground injection			
	Discharge to Publicly Owned Treatment Works (POTW)			
	Total Released to Water			
Land	On-site disposal			
	Off-site disposal			
	Total Released to Land			
Recycling and Treatment	On-site Recycling			
	Off-site Recycling			
	On-site Treatment			
	Off-site Treatment			
Source of Data				

Plant Visit Information	
Plant: _____	**Worksheet 4**
Date Of Visit: _____	*Corporate*
Company Representatives: _____	*Commitment*

Question	Company Response
A. Source Reduction Policy	
i. Does your company have a working definition of source reduction? What is it? When was it first established?	
ii. Does your company have a written policy favoring source reduction as the most desirable waste management option? If so, may we have a copy? Who establishes the policy?	
iii. Is the policy corporation-wide or plant-specific?	
iv. Is your source reduction program a multimedia program? That is, does it cover chemical releases to air, water, and land?	
v. Do you have specific source reduction plans for specific chemicals?	
vi. For specific industrial processes?	
vii. Does your plan include specific source reduction goals? How are they measured?	
viii. Are source reduction options considered during the engineering phase when planning new product lines?	

Question	Company Response
B. Company Leadership	
i. Is there an individual responsible for source reduction within your plant? If so, what is that individual's title and technical/program management background?	
ii. Is there a department or division at the company or plant that is responsible for source reduction? If so, how is it staffed, where is it located, and what are its general responsibilities?	
iii. How do this division and individual fit into the overall corporate structure? That is, to whom do they report and who reports to them?	
iv. To what extent are the managers and operators of your production processes involved in source reduction? For example, do they identify opportunities, implement changes, or measure progress? (Research shows that many source reduction opportunities are identified by production personnel.)	
C. Source Reduction Incentives	
i. Is there an incentive system that encourages employees to come up with suggestions for source reduction? What are the incentives?	
ii. Is there a source reduction training program? If so, please describe.	
iii. Is source reduction information transmitted throughout the plant/company in the form of newsletters/staff meetings? May we see written examples?	

Plant Visit Information

Plant: _____

Date Of Visit: _____

Company Representatives: _____

Question	Facility Level (facility-level materials accounting)	Process Level (process-level source reduction inventory)
A. Facility and Process Level Data		
i. Does the company collect waste-related data at this plant?		
ii. How are materials tracked (as specific chemicals, by waste categories, as products, etc.)?		
iii. Which chemicals and/or waste categories are tracked?		
iv. Does the company track materials released to all environmental media (air, land, and water)?		
v. Are materials tracked in each environmental medium regardless of whether they are regulated in them?		
vi. Are materials tracked in both production and non-production areas of the plant? (Nonproduction areas include storage, loading, transfer, and pollution control sites.)		
vii. Are chemical-specific materials balances performed to assure identification of all major sources and quantities of waste generated?		

Question	Facility Level (facility-level materials accounting)	Process Level (process-level source reduction inventory)
viii. If a materials balance is not performed, why not?		
ix. Are the nature and levels of uncertainty in the measurement methods used identified?		
x. How are these levels of uncertainty factored into the analysis of the overall results? (Make sure you get some numbers on the amount of material not captured by the measurement methods used.)		
xi. Who sees the results of the data collection program?		
xii. What are the data used for?		
xiii. How do these data factor into evaluations of both the source reduction policy and the specific actions taken to achieve reductions in waste generation?		
xiv. How frequently are the data collected?		
xv. Who performs the collection procedure? What is this person's educational and professional background?		

Question	Facility Level (facility-level materials accounting)	Process Level (process-level source reduction inventory)
xvi. Have you computerized this information?		
xvii. Have you established a base year from which present and future accomplishments are measured? If so, what is that year?		

Question	Company Response
B. Progress Reports	
i. Does the plant prepare progress reports?	
ii. How do you measure source reduction accomplishments (e.g., by pounds per unit of production)?	
C. Full Cost Accounting	
i. Are waste-related costs accounted back to their source in the company's cost accounting system? Or does the company treat environmental costs as a fixed overhead expense?	
ii. Are wastes released to all environmental media included in the cost accounting?	
iii. Are costs designated by specific chemicals, chemical categories, or waste categories?	

Question	Company Response
iv. What types of costs are captured by the full cost accounting method you use:	
a. Materials (e.g., the costs of wasted starting materials and lost products)?	
b. Waste handling (e.g., capital and operational expenses for on-site recycling, treatment, storage, or disposal facilities; transportation and other expenses for wastes sent off-site)?	
c. Regulatory compliance?	
d. Insurance?	
e. Future liabilities from waste generation (e.g., accidents, worker illness, or waste site clean-up)?	
f. Public and customer relations dealing with waste issues?	
g. Other (please specify)?	

Plant Visit Information

Plant: _____

Date Of Visit: _____

Company Representatives: _____

Achieved or Planned			
Specific Waste Reduced (hazardous or nonhazardous)			
Waste Medium			
Source Reduction Type (see code below)*			
Source Reduction Practice			
Percent Waste Reduced			
Amount Waste Reduced			
Motivation			
How Identified			
Comments			
OPTIONAL	Change in Yield			
	Dollars Saved			
	Dollars Spent			

*Source reduction types: CS, chemical substitution; EQ, equipment change; OP, operational change; PR, product reformulation; PS, process change.

Plant Visit Information

Plant: _____

Date Of Visit: _____

Company Representatives:_____

A. Establishing Priorities

Please indicate the order of importance of each of the following factors in terms of its influence on your source reduction program.

☐	Reducing releases of the most hazardous chemicals
☐	Complying with regulations
☐	Avoiding potential future liability
☐	Saving raw material costs or otherwise improving process efficiency
☐	Avoiding waste management costs
☐	Other (please specify)

B. Incentives for Source Reduction

Please identify which incentives influenced your plant's decision to implement source reduction, and indicate their relative importance on a scale of 1-10 (10 being most important).

i.   The Toxics Release Inventory required by EPA, and the fact that it will be made public, create a strong incentive for companies to implement source reduction policies.

   Yes ☐   No ☐   Importance (1-10) ☐

ii. Source reduction makes economic sense because it can save raw materials and help to make production more cost-effective.

   Yes ☐   No ☐   Importance (1-10) ☐

iii. The state has provided us with information or technical assistance that helped us to reduce waste at the source.

   Yes ☐   No ☐   Importance (1-10) ☐

iv. Trade associations or other industry sources provided us with information or technical assistance that helped us to reduce waste at the source.

   Yes ☐   No ☐   Importance (1-10) ☐

v. Increasingly stringent regulations have made it necessary to find innovative ways to reduce the cost of compliance.

Yes [   ]   No [   ]   Importance (1-10) [   ]

vi. The state (or local government) has persuaded us to put a source reduction program into place as a result of an enforcement action or negotiation over permit conditions.

Yes [   ]   No [   ]   Importance (1-10) [   ]

vii. We look for ways to reduce pollutants that are not currently regulated to cut down on the need for future regulation or to avoid the need for compliance costs if such materials are regulated in the future.

Yes [   ]   No [   ]   Importance (1-10) [   ]

C. Barriers to Source Reduction

Please identify those statements with which you agree and indicate their relative importance on a scale of 1-10.

i.   We have experienced significant engineering or technical barriers to source reduction.

Yes [   ]   No [   ]   Importance (1-10) [   ]

ii. We have experienced cost barriers to source reduction (please specify).

Yes [   ]   No [   ]   Importance (1-10) [   ]   [   ]

iii. We have identified serious regulatory obstacles to source reduction (please specify).

Yes [   ]   No [   ]   Importance (1-10) [   ]   [   ]

iv. We could benefit from state technical assistance.

Yes [   ]   No [   ]   Importance (1-10) [   ]

Recent INFORM Publications

To order any of the following publications, please indicate below the number ordered. Substantial discounts for 5 or more — please write for information. All orders must be prepaid. Please fill out both sides of this form.

CHEMICAL HAZARDS PREVENTION

___ **Promoting Hazardous Waste Reduction: Six Steps States Can Take** (Warren R. Muir, Ph.D., and Joanna D. Underwood), 1987, 21 pp., $3.50

___ **Cutting Chemical Wastes: What 29 Organic Chemical Plants are Doing to Reduce Hazardous Wastes** (David J. Sarokin, Warren R. Muir, Ph.D., Catherine G. Miller, Ph.D., and Sebastian R. Sperber), 1986, 535 pp., $47.50

___ **Tracing a River's Toxic Pollution: A Case Study of the Hudson** Phase I, 1985, 150 pp.; Phase II, 1987, 209 pp., (Steven O. Rohmann, Ph.D.), Set: $20.00

___ **Preventing Pollution Through Technical Assistance: One State's Experience** (Mark H. Dorfman and John Riggio) 1990, 72 pp., $15.00

___ **Toxics in Our Air** (Nancy Lilienthal), 1990, 8 pp., $4.50

___ **Trading Toxics Across State Lines** (Nancy Lilienthal) 1990, 32 pp., $7.50

___ **Toxic Clusters: Patterns of Pollution in the Midwest** 1991, 106 pp., $15.00

MUNICIPAL SOLID WASTE

___ **Burning Garbage in the US: Practice vs. State of the Art** (Marjorie J. Clarke, Maarten de Kadt, Ph.D., and David Saphire), 1991, 275 pp., $47.00

___ **Business Recycling Manual** (copublished with Recourse Systems, Inc.), 1991, 202 pp., $85.00. (Please add $5.00 for shipping)

___ **Garbage Management in Japan: Leading the Way** (Allen Hershkowitz, Ph.D., and Eugene Salerni, Ph.D.), 1987, 130 pp., $15.00

___ **Garbage Burning: Lessons from Europe: Consensus and Controversy in Four European States** (Allen Hershkowitz, Ph.D.), 1986, 53 pp., $9.95

___ **Technologies for Minimizing the Emission of NO_x from MSW Incinerators** (Marjorie J. Clarke), 1989, 33 pp., $9.95

___ **Improving Environmental Performance of MSW Incinerators** (Marjorie J. Clarke), 1988, 82 pp., $15.00

___ **Solid Waste Management: The Garbage Challenge for New York City** (Maarten de Kadt, Ph.D., and Nancy Lilienthal), 1989, 56 pp. $7.95

(Continued on Next Page)

119

(Order form continued from previous page)

OTHER

__*Controlling Acid Rain:*
A New View of Responsibility
(James S. Cannon), 1987, 55 pp., $9.95

__*Winning with Water:*
Soil-Moisture Monitoring for Efficient
Irrigation
(Gail Richardson, Ph.D., and Peter Mueller-
Beilschmidt, P.E.), 1988, 192 pp., $24.95

__*Drive for Clean Air:*
Natural Gas and Methanol Vehicles
(James S. Cannon), 1989, 252 pp., $65.00

Subtotal $_____

Postage and handling _____
(see "Shipping Fees," right)

Total $_____

Enclosed is my check for $_____

Forthcoming Publications

**Environmental Dividends: Cutting
Chemical Wastes 1991**

**Tackling Toxics in Everyday Products:
A Directory of Organizations**

**Source Reduction Planning for
Municipalities**

Reducing Office Paper Waste

*Prices and approximate dates of release of
forthcoming publications are available upon
request.*

Shipping Fees

Please add handling charges as follows:

US (4th class delivery; allow 4-6 weeks);
add $3 for first book + $1 each add'l. book

Canada: add $5 for first book + $3 for each
add'l. book

Priority shipping is higher; please call for
charges

For orders over $10.00 only: ❑ VISA   ❑ MASTER CARD   ❑ AMEX
Card # _____ Expiration Date _____
Signature _____ Phone # _____

Name_____ Title _____

Company/Affiliation _____

Address _____

City _____ State _____ Zip _____

❑ I would like to support INFORM. A tax-deductible contribution of $25 or more entitles me to
a year's subscription to INFORM's quarterly newsletter, *INFORM Reports.*

Please make checks payable to INFORM. Mail to INFORM, 381 Park Avenue South,
New York, NY 10016-8806 (212) 689-4040.

About the Author

Lauren Kenworthy

Lauren Kenworthy is an environmental consultant who previously served as legislative assistant to United States Representative Howard E. Wolpe (Democrat, Michigan), specializing in environmental issues. Ms. Kenworthy drafted the Waste Reduction Act, which was introduced in Congress in 1986 by Representative Wolpe. She was also responsible for educating congressional staff and members of the public about issues and options for promoting source reduction of industrial toxic wastes. Before working on Capitol Hill, Ms. Kenworthy was a research assistant at the Natural Resources Defense Council.

Ms. Kenworthy is currently studying for her Ph.D. at the University of Maryland. She received her bachelor's degree with honors from Yale University.

INFORM Board of Directors

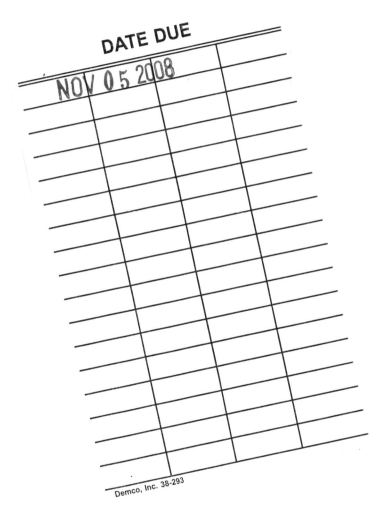

DATE DUE

NOV 0 5 2008